FIRE PREVENTION
Inspection and Code Enforcement
4th Edition

David Diamantes

Copyright © 2015 David Diamantes

All rights reserved

ISBN: 1508481172
ISBN-13: 978-1508481171

David Diamantes

DEDICATION

To my wife Bonita, who picks me up when I fall, carries me when things get tough, supports me in all my endeavors, and defends me even when I'm wrong.

ALL RIGHTS RESERVED. No part of this work covered by the copyright hereon may be reproduced in any form or by any means—graphic, electronic, or mechanical, including photocopying, recording, taping, Web distribution, or information storage and retrieval systems—without the written permission of the publisher.

PREFACE

WHY FIRE PREVENTION?

On February 20, 2003, foam plastic soundproofing material was ignited by an indoor pyrotechnic display at the Station nightclub in West Warwick, Rhode Island. The fire claimed one hundred lives and resulted in two hundred injuries. The press had a heyday. The public was outraged. Elected officials questioned the safety of public assemblies within their jurisdictions. Fire officials scrambled to provide answers to all the questions while checking to make sure that nightclubs in their own jurisdictions were safe. Unfortunately, it takes a catastrophe to remind the public that the danger from fire is real and that fire affects real people, not just names in the newspaper. But the loss of life isn't the only tragedy caused by fires.

The incidence of fire in the United States says a lot about us as a people. Study after study has identified the loss from fire in our country as a national disgrace, yet the American people, the media, and all levels of government are unaware of or unable to recognize that a problem even exists. Many in the fire service are loath to dedicate staff and resources to the very effort that led to the development of the municipal fire service in this country—fire prevention. Many in fire service management view fire prevention as a political liability. Some have even gladly relegated the function to another agency or, worse, abandoned the effort in favor of other functions. The image of a fire inspector armed with a code book and a clipboard rarely makes the news unless it involves a politically sensitive issue.

The roots of today's fire service are anchored in the ashes of the great conflagrations of the late nineteenth and early twentieth-centuries when fire brigades and salvage companies were commissioned by the stock fire insurance companies for the protection of insured properties. Perhaps we have forgotten where we came from. The protection of life and the preservation of property through education, regulation, and enforcement were and still are the jobs of the fire service. The very existence of many communities revolves around a few select industries. Most of the citizenry is employed within those industries, by their suppliers, or within the service sector that provides housing and basic goods and services.

When fire destroyed Farmland Foods' processing center in Albert Lea, Minnesota, on July 9, 2001, seven hundred of Albert Lea's citizens were thrown out of work at the city's second largest employer.[1] The unemployment rate literally doubled overnight. The effect on the region's economy will be severe at best—devastating at worst. Mortgage and car loans will default; stores and restaurants will suffer a decline in business as belts are tightened. The economic trickle-down impacting the lives of literally thousands of people can be traced back to sparks from a welder's torch that ignited combustible materials. One of the most common and easily prevented causes of industrial fires has again changed the future of an entire region.

If a method to prevent earthquakes, tornadoes, or hurricanes were discovered, the public would demand swift implementation. Government and industry would devote millions of dollars and dedicate the best and the brightest to the project. Compare the pipe dream of controlling the weather with the reality of America's fire record, and the results are sobering:

Fire in the United States, Eleventh Edition, August 1999

The fire problem in the United States, on a per capita basis, is one of the worst in the industrial world. Thousands of Americans die each year in fires, tens of thousands of people are injured, and property losses reach billions of dollars. To put this in context, the annual losses from floods, hurricanes, tornadoes, earthquakes, and other natural disasters combined in the United States average just a fraction of the losses from fires. The public in general, the media, and local governments, however, are generally unaware of the magnitude and seriousness of fire to the communities and to the country.[2]

Recommissioned Panel for America Burning, May 2000

The frequency and severity of fires in America do not result from a lack of knowledge of the causes, means of prevention or methods of suppression. We have a fire "problem" because our nation has failed to adequately apply and fund known loss reduction strategies.[3]

Recommissioned Panel for America Burning, May 2000

America today has the highest fire losses in terms of both frequency and total losses of any modern technological society. Losses from fire at the high rate experienced in America are avoidable and should be as unacceptable as losses caused by drunk driving or deaths of children accidentally killed playing with guns.[4]

National Association of Credit Men, Committee on Fire Insurance, June 1909

Too frequently an exhaustive report on (fire safety) conditions is treated by the municipal authorities with an indifference akin to contempt.[5]

When I began my career as a firefighter, it never occurred to me that fires were not only a threat to my community, but a threat to my livelihood as well. I reasoned that the higher the fire frequency and the greater the damage, the more the public would realize how vital my services were to the community. I could not have been more wrong. Every fire affects every member of the community through increased costs for government services, loss of jobs and productivity, increased insurance premiums, and erosion of the tax base. The municipal budget that funds the fire department is the same one that funds every other local government service and is entirely dependent on the local tax base. When forced to choose between government services, the public chooses education first. Who can resist the mantra "It's for the children"? Parks, libraries, public safety, and other services vie for whatever is left. Fire is always something that happens to other people. When forced to choose between a service they feel directly affects them and those that only affect other people, which will the public choose?

The process of inspection and the enforcement of the fire code is one part (albeit an important one) in the fight against one of our nation's most significant problems. You have the means to play a part in an endeavor with a greater potential economic impact than controlling the weather. If you do not believe in preventing fires for the good of your community, believe in it for your own job security.

HOW TO USE THIS BOOK

Throughout this text, the model codes and referenced standards are described as the tools of the trade for the fire inspector. Learning to be an inspector is, in fact, mostly a lesson in how to use these tools. Using this book without access to a model building and fire code is like studying a how-to manual on fly-fishing without having a fly rod. The greatest challenge in writing the original text and in updating this fourth edition was attempting to describe the basic principles of the fire inspection process within the different code systems in use today. Much of the first chapter, on code administration, is devoted to describing the different approaches taken by local, state, and federal government in adopting building and fire prevention regulations.

There are now two model code systems in use within the United States. The "big three" regional model code organizations, Building Officials and Code Administrators International (BOCA), International Conference of Building Officials (ICBO), and Southern Building Code Congress International (SBCCI), abandoned their respective model codes and cooperatively developed the International Codes in 2000. They got along so well that the three groups consolidated in 2003 and exist today as the International Code Council. As of February 2015 the *International Building Code* (IBC) has been adopted by or within all fifty states, Washington, DC, and the Department of Defense. The *International Fire Code* (IFC) has been adopted by or within forty-two states and Washington, DC.

The National Fire Protection Association (NFPA) publishes NFPA 5000, *Building Construction and Safety Code* and NFPA 1, *Uniform Fire Code*, in addition to almost three hundred other codes standards and recommended practices. NFPA 5000 has not been widely adopted. The *Uniform Fire Code* (formerly printed by ICBO for the Western Fire Chiefs Association) is used by several states and numerous jurisdictions.

As an inspector, you will not have a much of a say in the code system adopted in your jurisdiction anymore than you get to pick the mayor or city council. You will work with the tools they give you, and attempt to provide the best service to the public and the business community you can. In the aftermath of a fire at the Imperial Foods processing plant in Hamlet, North Carolina, in 1991, in which twenty-five were killed and fifty-four injured, the local fire chief stated that the entire incident centered around one problem—lack of enforcement of existing codes.[6] What code is not as important as consistent, competent, and evenhanded inspection and enforcement.

FEATURES OF THIS BOOK

There are many features contained within the chapters of this book, which will assist you in learning the valuable concepts associated with fire prevention.

- Safety is emphasized throughout the book, including a chapter on general fire safety provisions (9), to ensure that the proper precautions are utilized to help prevent fire and loss of life.
- Key terms and notes highlighted in the margins of each chapter reinforce the critical knowledge required for effective performance on the job.
- Review questions provide an important review of the content in each chapter, while Discussion questions present scenarios to strengthen effective identification of hazards and efficient problem-solving skills.
- FESHE correlation to the National Fire Academy course on fire prevention illustrates how and where the content in the book meets the desired outcomes for this course, in order to promote standardized training.
- Comprehensive appendices in the back of the book include important mathematical references and a Glossary to provide an informational guide that is handy for use in the classroom or on the job.

NEW TO THIS EDITION

- *International Building Code* (IBC), *International Fire Code* (IFC) and *Uniform Fire Code* (UFC) 2012 references, including code tables, provide an up-to-date explanation of the rules and regulations behind building code safety.
- Organization and effective use of the IBC, IFC, and UFC are covered to ensure a clear understanding of these important standards.
- New inspection requirements based on the advancement of technology are covered in order to keep pace with the industry.
- A report on recent fire events reinforces the necessity of fire prevention and how these events influence changes in the international code-making process.

SUPPLEMENT TO THIS BOOK

An e-resource CD-ROM is available for instructors, and includes many handy tools for effective classroom presentation and student evaluation.

- An Instructor's Guide includes lesson plans with PPT correlations, as well as answers to review and discussion questions.
- A PowerPoint® presentation for each chapter highlights the critical points and enhances classroom lectures.
- A correlation to NFPA Standard 1031 and the FESHE course on Fire Prevention provides the required references for instructor preparation of course content.

David Diamantes

Fire and Emergency Services Higher Education (FESHE)

In June 2001, The U.S. Fire Administration hosted the third annual Fire and Emergency Services Higher Education Conference, at the National Fire Academy campus, in Emmitsburg, Maryland. Attendees from state and local fire service training agencies, as well as colleges and universities with fire related degree programs attended the conference and participated in work groups. Among the significant outcomes of the working groups was the development of standard titles, outcomes, and descriptions for six, core associate-level courses, for the model fire science curriculum that had been developed by the group the previous year. The six core courses are Building Construction for Fire Prevention, Fire Behavior and Combustion, Fire Prevention, Fire Protection Systems, Principles of Emergency Services, Principles of Fire and Emergency Services Safety and Survival. [1] Since that beginning, committees have continued to refine and improve the model curriculum. This text was written around the desired outcomes of *Fire Prevention* and *Principles of Code Enforcement*.

Fire Prevention Course Content

The National Fire Science Curriculum Advisory Committee identified eight desired outcomes for the course. This text was written to address each outcome.[2]

Desired Outcomes

1. Define the national fire problem and role of fire prevention.
2. Identify and describe fire prevention organizations and associations.
3. Define laws, rules, regulations, and codes and identify those relevant to fire prevention of the authority having jurisdiction.
4. Define the functions of a fire prevention bureau.
5. Describe inspection practices and procedures.
6. Identify and describe the standards for professional qualifications for Fire Marshal, Plans Examiner, Fire Inspector, Fire and Life Safety Educator, and Fire Investigator.
7. List opportunities in professional development for fire prevention personnel.

8. Describe the history and philosophy of fire prevention.

Principles of Code Enforcement Content

Desired Outcomes:

1. Explain the code enforcement system and the fire inspector's role in that system.
2. Describe the codes and standards development and adoption processes.
3. Describe the difference between prescriptive and performance-based codes.
4. Describe the legal authority and limitations relevant to fire code inspections.
5. Describe the importance of thorough documentation.
6. Recognize ethical practices for the code enforcement officer.
7. Explain the application, and interrelationship of codes, standards, recommended practices and guides.
8. Describe the differences in how codes apply to new and existing structure.
9. Indentify appropriate codes and their relationship to other requirements for the built environment.
10. Describe the political, business, and other interest that influence the code enforcement process.
11. Identify the professional development process for code enforcement practitioners

An NFPA 1031 JPR matrix is included at the end of the text which lists each JPR for Inspector I and II and the corresponding chapter(s) that address it.

CONTENTS

1	Code Administration	5
2	Inspection	23
3	Use and Occupancy	51
4	Building Limitation and Types of Construction	67
5	Fire Resistive Construction Elements	87
6	Installation of Fire Protection Systems	107
7	Means of Egress	141
8	Interior Finish Requirements	169
9	General Fire Safety Provisions	177
10	Maintenance of Fire Protection Systems	201
11	Means of Egress Maintenance and Fire Evacuation Planning	225
12	Hazardous Materials	245
13	Flammable Liquids and Aerosols	259
14	Detonation and Deflagration Hazards	279
15	Hazardous Assembly Occupancies	305
16	Other Storage and Processing Occupancies	313
17	Compressed Gases and Cryogenic Fluids	323
18	Pesticides and Other Health Hazards	337
19	Using Referenced Standards	343
	Appendix A - Conversion Charts and Related Mathematics	347
	Notes	355

Glossary — 363

NFPA 1031 Matrix — 371

ACKNOWLEDGMENTS

I want to thank Cengage Publishing for making this 4th edition possible. They graciously returned my publishing rights when they discontinued their fire science line. It was a kind gesture they didn't have to make. To their management and to the editors I worked with for so many years, thank you. My heartfelt thanks go to Sandy Clark, Alison Weintraub, Jennifer Star, Deborah Yarnell, Erin Coffin, Patty Garrison, Patty Stephen, Larry Main, Jennifer Hanley, Linda Verde Conto, Dawn Daugherty, Fair Huntoon, Maura Theriault, Brian McGrath, Mary Ellen Black, Ruth Fisher and Janet Maker. To anyone I have missed I beg your forgiveness. A special thanks goes to Mark Huth, who called me one day out of the blue and asked if I wanted to write a book.

I also want to thank the fire service professionals who reviewed previous editions and set me straight when I needed it. Special thanks goes to my A-team who reviewed this 4th, self-published edition. They are:

- Gary D. Kistner, Program Coordinator/Director of Graduate Studies Fire Service Management Southern Illinois University Carbondale
- A. Maurice Jones, Senior Fire Protection Engineer, City of Alexandria, Virginia
- Chief Ken Van Sparrentak, CFO/EFO, Homeland Security, Fire Fighter Technology, Oakland Community College, MI
- Cynthia Wong, Fire Specialist II, Carlsbad, California Fire Department
- Chief Russ Chandler, Virginia Fire Marshal Academy (retired)
- Fire Marshal Todd Iaeger, Town of Chapel Hill, North Carolina
- Chief John Senft, York, PA, Department of Fire/ Rescue Services, (retired)
- Ronald F. Hoelle II, USAF Ret, Fire Protection, Department of Defense
- Jan Mitchell, MPA, State of Alaska Fire Training Administrator

David Diamantes

Section 1

FIRE PREVENTION THROUGH REGULATION

Fire prevention inspection is no longer the exclusive bailiwick of uniformed firefighters assigned to fire prevention bureaus. Today's inspector is likely to be a civilian combination inspector who performs building inspections, property maintenance inspections, and fire inspections, all under the authority of the fire chief. Every conceivable combination of uniform/civilian staffing, enforcement agency organization and division of responsibility has been tried or is in use today in the United States. Tight budgets and the philosophy of making government leaner and smarter drive the efforts.

A few fire prevention offices have the sole responsibility of enforcing the fire prevention code. Others investigate fires and have law enforcement responsibilities. Still others are involved in plan review, acceptance testing, and the inspection of new construction. The arrangement with the most potential significant impact is the office that combines the traditional responsibilities of the building department, fire inspection office, and fire investigation office under one director within the fire department.

The benefits of a coordinated effort include increased public safety and economic attractiveness for the jurisdiction. When fire department operational effectiveness and firefighter safety are considered from the planning stage through development, with constant feedback through the investigation process, everyone wins including the business community. With the mix of roles and responsibilities within many jurisdictions, legal authority can sometimes become a confusing issue. Many of us wear many hats, such as fire marshal, fire inspector, fire investigator, building inspector, or a combination of some or all of the above. These roles may exist in statute or regulation, and may overlap. Always remember which hat you have on: You have only the authority and responsibility of your position.

Chapter 1
Code Administration

Learning Objectives

Upon completion of this chapter, you should be able to:
- Describe the model code process and name the model code organizations.
- Describe the origin of the first National Building Code.
- Describe the code adoption process used by state and local governments and explain the basis for their authority to enact such regulations.
- Describe the economic forces behind the development of a single national model code.

INTRODUCTION

Essential qualifications for the successful Inspector are ability to recognize fire hazards when he sees them, a logical and practical understanding of how to safeguard them, persistence in securing necessary corrections, and absolutely impartial and accurate observation and reporting of conditions found.[1]

(From *Field Practice—An Inspection Manual for Property Owners, Fire Departments and Inspection Offices*, 1922, National Fire Protection Association.)

Fire prevention as practiced by the Romans in the first century was quite different from the system of technical **codes** that has evolved and is used throughout most of the industrialized world. The "code" enforced by the Vigiles, a seven-thousand-man fire department, was short and sweet: Have a fire? Get punished. The first-ever building regulations, decreed by King Hammurabi, founder of the Babylonian empire, were equally succinct. *If the building collapses and kills the owner, the builder shall be put to death. If the owner's son is killed, the builder's son shall be put to death.*

These early societies regulated behavior by ensuring punishment in the event of failures. They specified the failure to be avoided and the punishment to be meted out, but they did not specify the procedures to be followed in order to prevent the failure. Technical codes, which specified construction methods and materials, requirements for material handling, and regulations for certain processes, came much later, but not until after centuries of conflagrations and at a cost of countless lives and dollars.

NOTE

Early societies regulated behavior by ensuring punishment in the event of failures.

The statement that our fire and building codes are written with a blood-soaked quill dipped in ashes isn't far off the mark. They were born from the conflagrations that have plagued the United States during much of our history. In Joseph K. Freitag's 1917 treatise *Fire Prevention and Protection as Applied to Building Construction*, he cited the conflagrations in Table 1-1 as evidence that fire prevention and protection weren't local problems.

"All large cities contain localities which are pregnant with conflagration possibilities, principally due to the rapid, haphazard growth and construction of such cities. Large areas of wooden buildings may exist, as in San Francisco; or a large store or warehouse, stocked with inflammable goods, inadequately safeguarded, as at Baltimore may provide the cause. The absence of fire walls, shutters or window protection may turn an ordinary fire into one of great magnitude; while such circumstances as low-water pressure, delay in transmitting alarms, bad judgment or disorganization of the fire department, have all been responsible for wide-spread fires."[2]

Conflagrations in the United States

Date	City	Fire Loss
June 1820	Savannah, Georgia	$3,000,000
December 1835	New York City	$17,000,000
December 1835	Charleston, South Carolina	$6,000,000
March 1852	New Orleans	$5,000,000
July 1866	Portland, Maine	$10,000,000
October 1871	Chicago (56 insurance companies bankrupted)	$168,000,000
November 1872	Boston (65 acres of the city)	$70,000,000
June 1889	Seattle	$5,000,000
November 1889	Lynn, Massachusetts	$5,000,000
October 1892	Milwaukee	$5,000,000
May 1901	Jacksonville, Florida	$11,000,000
February 1904	Baltimore (140 acres, 1343 buildings)	$40,000,000
April 1906	San Francisco (earthquake and fire)	$350,000,000
April 1908	Chelsea, Massachusetts (3500 buildings)	$12,000,000

Table 1.1 Source: Joseph K. Freitag, *Fire Prevention and Protection as Applied to Building Construction*, Wiley, New York, 1912, p.7.

In the aftermath of disastrous fires, politicians passed sweeping laws aimed at preventing future conflagrations. Perhaps the most famous conflagration in American history occurred in Chicago, in October 1871. Over 17,500 buildings were destroyed, over 300 persons were killed, and over 100,000 were left homeless in the "Great Chicago Fire."[3]

Within a month of the fire, Joseph Medill, publisher of the *Chicago Tribune*, was elected Chicago's 23rd mayor. Medill ran as the "Fireproof Party" candidate, promising to keep Chicago from ever suffering the scourge of another conflagration.

His "Fireproof Party" label was proof that even a catastrophe like the Great Chicago Fire can be leveraged politically (Figure 1-1). Within three years of the fire, however, reports had reached the National Board of Fire Underwriters (an arm of the insurance industry) that fire safety conditions in Chicago were deplorable.

Figure 1-1. Joseph Medill, publisher of the Chicago Tribune and founder of the Fireproof party.

The Board's investigating committee's report on conditions in Chicago revealed that conditions were actually worse than before the fire in 1871. The fire department was neglected by the city commissioners and was ill trained and poorly equipped. Fire wardens, entrusted with construction inspections, made no effort to enforce the city's building regulations.[4] The insurance industry literally strong-armed Chicago's elected officials into improving the fire department and fire safety conditions by threatening to cancel every insurance policy within the city. The city was given three months to comply. One week before the deadline, the city conceded at the insistence of the business community, who clearly understood the potential impact of losing insurance coverage.

If any city had good reason to enact and enforce fire and building codes in the aftermath of a tragedy, it was Chicago. But the windy city's experience of adopting and then neglecting regulations was typical of many communities. As the signs of the devastation were wiped away and the city was rebuilt, memories faded and regulations were no longer convenient or politically expedient. When the political system proved to be ill suited to develop effective fire prevention regulations, industry stepped up to the plate.

MODEL CODES

In 1905, the National Board of Fire Underwriters (NBFU), which later became the American Insurance Association or AIA, published the first *National Building Code*.[5] Unlike today's **model codes**, the *National Building Code* was developed by staff of the NBFU, not through the use of technical committees and the consensus process. The *National Building Code* was published by the NBFU through 14 editions over 75 years. The *National Building Code* pioneered the fundamental principles that drive our buildings codes today, such as building height and area restrictions, fire-resistive construction, shaft enclosures, fire-walls and doors, fire-stopping, means of egress, and fire suppression systems.[6]

NOTE

Unlike today's model codes, the National Building Code was developed by staff of the NBFU, not through the use of technical committees and the consensus process.

What was most significant about NFBU's action was that the *National Building Code* was made available for state and local governments to **adopt** at no charge. Governments were given their first opportunity to adopt a building code developed by experts from outside their political jurisdictions, free from local political influences.

The NBFU also developed a fire prevention code that was intended to be a companion document. The *National Building Code* was discontinued by AIA after 1976.[7]

NOTE

The National Building Code was made available for state and local governments to adopt at no charge.

Model Code Organizations

The *National Building Code* was developed by the insurance industry as a method of reducing conflagration potential. It was a property protection code that afforded an increased level of life safety by reducing the size and frequency of major fires. But make no mistake—the code was developed to protect the financial interests of the insurance industry.

In 1922, the Pacific Coast Building Officials Conference developed the first of three regional construction codes in the United States. Southern Building Code Congress International and the Building Officials Conference of America developed their codes in 1945 and 1950 respectively. These codes were initially developed by building officials organizations with the intention of improving the level of life safety while providing property protection. By 2000, the regional groups had discontinued their own codes and were cooperatively developing the International Codes, a single set of codes for the entire United States.

The International Code Council

On December 9, 1994, the International Code Council (ICC) was officially organized as a nonprofit umbrella organization consisting of representatives of BOCA, International Conference of Building Officials (ICBO), and Southern Building Code Congress

International (SBCCI). The sole purpose of the ICC was to facilitate the development and maintenance of the International Codes and to coordinate related supporting activities.[8]

The 1990s ushered in a series of political developments that brought our regional system of model code development under close scrutiny. The North American Free Trade Agreement (NAFTA) and actions with the European Common Market that effectively eliminated many barriers to European trade were cause for growing concern. Was our system of regional model codes with technical disparities between each a threat to American competitiveness? The response from the three model code groups was to form the ICC and begin development of a single set of codes. By 1996, the *International Plumbing, Mechanical,* and *Private Sewage Disposal Codes* were published. Under the agreement establishing the ICC, the members' groups agreed to discontinue publishing their individual codes as of 2000.

The Legacy Codes

When the three regional model code organizations consolidated their staffs, resources, and memberships as the International Code Council in 2003, the regional organizations disappeared, but their old codes didn't. Why should we even discuss the discontinued codes if they are no longer in use? *Because they ARE still in use.* The majority of the building stock in the United States was constructed prior to 2000, under a specific edition of one of the regional model codes. Until they are renovated or undergo a change in use, they must be maintained to the level of protection required by the code under which they were constructed. The codes developed by the regional organizations are referred to as legacy codes, and the regional organizations as legacy code organizations

> **NOTE**
>
> **The majority of the building stock in the United States was constructed under a specific edition of one of the regional model codes. Until they are renovated or undergo a change in use, they must be maintained to the level of protection required by the code under which they were constructed.**

Figure 1-2 represents legacy code adoptions by state. The Uniform Codes were firmly entrenched in the West, BOCA/National in the Northeast, and Standard codes in the South. Local governments in Texas, Oklahoma, Kansas and Missouri adopted different codes. Wisconsin and New York adopted and maintained their own, state-developed codes.

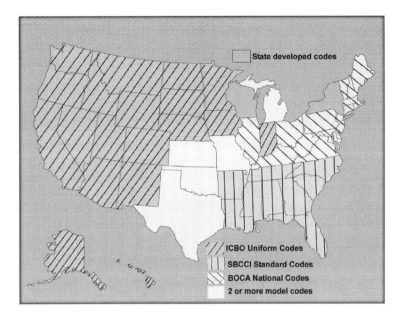

Figure 1-2. Model codes in place in the United States pre-2000

Building Officials and Code Administrators International (BOCA)

BOCA was established in 1915 by building officials from nine states and Canada as the Building Officials Conference of America,[9] "To discuss the principles underlying **ordinances** related to building."[10] In 1950, BOCA first published the *Basic Building Code*. The organization maintained building, mechanical, fire prevention, plumbing, and property maintenance codes through 1999. BOCA served the northeastern and Midwestern states.

International Conference of Building Officials

ICBO was established as the Pacific Building Officials Conference in 1921, and in 1927 published the *Uniform Building Code* (UBC), and numerous related codes that addressed housing maintenance, building security, outdoor signs, and building conservation.[11]

The *Uniform Fire Code* was maintained by the Western Fire Chief's Association and ICBO through the 1988 edition. The International Fire Code Institute (IFCI) was formed and published the codes since 1991. The *Uniform Mechanical Code* is maintained by the International Association of Plumbing and Mechanical Officials (IAPMO). ICBO served the western and Midwestern states.

Southern Building Code Congress International (SBCCI)

SBCCI was established in 1940 and in 1945 first published the *Standard Building Code*

(SBC).[12] In cooperation with the Southeastern Association of Fire Chiefs, SBCCI published the *Standard Fire Prevention Code* through 1999. Like BOCA and ICBO, SBCCI developed and maintained numerous related codes that addressed housing maintenance, building security, outdoor signs, and building conservation. SBCCI served the southeast.

National Fire Protection Association (NFPA)

NFPA is not a legacy model code organization, but is discussed here because the organization's key role in the legacy codes and the ICC process. NFPA was organized in 1896 and incorporated in 1930. There were five-distinct electrical codes in the United States, as well as nine radically different **standards** for sprinkler pipe size and spacing within 100 miles of Boston.[13] These conditions were instrumental in steering fire underwriters into forming an association to work toward uniformity in standards for fire protection. The group was to be known as the National Fire Protection Association.

By 1904, NFPA's active membership included 38 stock fire insurance boards and 417 individuals, most of whom were from the insurance industry. The first fire department officer to join NFPA was Battalion Chief W. T. Beggin of the Fire Department of New York (FDNY), who became a member in 1905. H. D. Davis, state fire marshal of Ohio joined the same year. In 1911, NFPA began maintaining the *National Electrical Code* and has published it ever since.[14]

Perhaps NFPA's greatest contribution to public safety is the standards they maintain. In simple terms, codes tell you what to do; standards tell you how to do it. The *International Building Code* requires hospitals and nursing homes to be protected with automatic fire sprinklers, in accordance with NFPA 13, *Standard for the Installation of Sprinkler Systems*. NFPA produces almost 300 codes, standards, and recommended practices developed by more than 205 technical committees. Among these, NFPA 70, *National Electrical Code*, in use throughout most of the United States, may well be the most widely used code of its type in the world.

NFPA 101, *Life Safety Code* addresses occupant safety in buildings with regard to the establishment and maintenance of exit facilities. It "does not attempt to address those general fire prevention or building construction features that are normally a function of fire prevention and building codes."[15] NFPA 101 separates buildings into occupancy types and then establishes requirements for new as well as existing buildings. The physical characteristics and limitations of the occupants as well as the structure are considered. This comprehensive standard addresses the building, occupants, and processes that occur. Many jurisdictions enforce the provisions of NFPA 101 in addition to their building and fire prevention codes. Often, certain provisions of NFPA 101 are more stringent and supersede those of the other codes.

In 2002, NFPA published NFPA 5000, *Building Code*. The code was developed as an alternative to the International Code Council's "I-Codes." NFPA partnered with the International Association of Plumbing and Mechanical Officials (IAPMO), American Society of Heating, Refrigerating and Air-Conditioning Engineers (ASHRAE) and the Western Fire Chiefs Association (WFCA) to offer a set of companion codes. NFPA 5000, the *Uniform Plumbing, Uniform Mechanical,* and *Uniform Fire Codes* are distributed by NFPA as the "C3" Codes. NFPA 5000 has not been widely accepted by state and local governments, however, the *Uniform Fire Code* is used in much of the western U.S. where the I-Codes are used as construction codes.

THE MODEL CODE PROCESS

The ICC code change process is based on those used in the past by BOCA, ICBO, and SBCCI, with a significant change in the voting process. In previous years, voting was limited to active code officials representing member government jurisdictions. The number of votes was based on the jurisdiction's population. Representatives of industry were permitted to propose changes and testify at hearings, but were not permitted to vote in the final action portion of the process. This was the fundamental difference between the ICC and NFPA processes. The intent was to limit voting to persons without a financial interest in the process. Those who disagreed with the system claimed that without industry votes, the codes were not "true consensus codes." Where the truth lies, only time will tell.

ICC Code Change Process

The ICC code change process is a three year cycle, with one-third of their 15 model codes being in a particular step at a time. Any person or interested group may submit a code change, regardless of affiliation with the ICC. They are editorially reviewed by staff, published, and distributed to the membership.

- Technical committees hear testimony on the proposed changes at public hearings at the national conferences of the model code groups. Committees are composed of code officials, industry representatives, members of academia, and special experts. Anyone may testify and participate at no cost. Committees approve, deny, or approve a modified form of each code change.
- Immediately after the vote by the technical committee, the moderator asks for a motion from the floor to challenge or overturn the committee action. All members of the ICC in attendance at the public hearing are eligible to vote on floor motions.
- The results of the public hearings are published and distributed to interested parties. This action is designed to give attendees at the upcoming annual meeting an opportunity to consider specific objections to the results of the public hearings and prepare testimony for the public comment process that precedes the final action hearing.

- Discussion during the public comment process is limited to items that meet the following criteria: items for which a public comment has been submitted; items which may result in a technical inconsistency between the *International Residential Code* and the associated ICC International Code, and items which received a successful assembly action at the public hearing.
- During final action consideration hearings, testimony is limited to code change proposals that have an assembly action or public comment and proposals on which the public hearing actions result in a technical inconsistency between the *International Residential Code* and the associated ICC International Code. Only code officials representing government entities are eligible to vote in the final action hearings.
- The final action on all proposed code changes is published as soon as practicable after the determination of final action. Code changes are published with the next edition of the code.

NFPA's Code Change Process

Where the ICC manages 15 codes, NFPA maintains over 200 codes, standards, and recommended practices. By necessity, their process relies on over 250 technical committees made up of over 8,000 volunteers. The process takes between 105 and 140 weeks and relies on meetings of the individual technical committees. NFPA membership is made up of 22% registered design professionals (architects and engineers): 5% representatives of business and industry; 12% representing health care facilities; 20% fire service; 3% insurance industry; 9% federal, state, and local government: 12% safety equipment manufacturers and distributors; 6% trade and professional associations; and 11% from other fields and disciplines.

Code change proposals are accepted from any interested individual or group. After discussion, the committees develop a draft, accepting, modifying or rejecting the proposal. The results are made available for public comment before a second round of committee meetings. Final action is taken by the membership at the association technical meeting.

The members of NFPA's technical committees can be found on the opening pages of every NFPA standard. The 1995 edition of NFPA 251, Standard Methods of Tests of Fire Endurance of Building Construction and Materials, lists 19 principals (voting members) 11 alternates, and two non-voting member. My unscientific assessment based solely on the roster is that the committee is composed of two fire service representatives, two representatives of the insurance industry, six representatives of various testing and research labs, a representative from a major university, and the remainder representing various trade associations and industry groups from forest products to steel to gypsum.[16]

NFPA ensures that no one group or groups are able to control a committee by limiting membership. NFPA's Standards Council and has adopted Regulations Governing Committee Projects specifically to keep the standards-making process fair and above board. Each committee member is classified according to guidelines into one of nine categories:[17]

1. Manufacturer
2. User
3. Installer/maintainer
4. Labor
5. Enforcing authority
6. Insurance
7. Special expert
8. Consumer
9. Applied research/Testing laboratory

Each committee is structured so that not more than one-third of the membership represents a single interest.

Is the fire service adequately represented in the process? Many in the fire service would emphatically answer no. Committee membership is an expensive proposition. Travel and lodging costs associated with committee meetings are difficult for most fire departments to justify to skeptical city managers and budget offices. And sometimes unfortunately, such skepticism exists within our own ranks. Some chiefs see little advantage in participating in the process. Industry recognizes the cost of committee participation for what it really is-- a legitimate business expense. Most local governments either lack the perspective or the resources.

AUTHORITY TO ENFORCE MODEL CODES

Rarely does a fire inspector ponder the source of his authority to inspect and order compliance as he drives between inspections. Most of an inspector's time between stops is filled with the more mundane world of hurried phone calls to the office, last minute inspection requests, and reports of inoperative fire protection systems and blocked exits. But the knowledge of the basis for that authority as well as the responsibilities that go with it should always be in the back of the inspector's mind.

NOTE

The legal basis for the promulgation and enforcement of building and fire prevention regulations by state and local governments actually originates within the Tenth Amendment to the United States Constitution.

The legal basis for the promulgation and enforcement of building and fire prevention regulations by state and local governments actually originates within the Tenth Amendment to the United States Constitution:

> *The powers not delegated to the United States by the Constitution, nor prohibited by it to the states, are reserved to the states respectively, or to the people.*

Police power, the fundamental power of the state to place restraints upon the personal freedom and property rights of individuals for the protection of the public health, safety, and welfare, is an inherent power of the states, possessed by them before adoption of the United States Constitution and reserved to them under the Tenth Amendment.[18]

Notice, however, that the Tenth Amendment made no mention of those powers reserved for municipalities. That is because there are none. The powers of local governments are only those powers delegated by the states. In the late 1800s, Chief Justice John Forrest Dillon of the Iowa Supreme Court wrote what was to become known as **Dillon's rule**. When arguments came before the Iowa Supreme Court regarding the powers of local governments, Justice Dillon's opinion became the rule for Iowa and later most of the states.

NOTE

Simply stated, under Dillon's rule, local governments have only those powers expressly conferred by the state constitution, state statutes, or home rule charter.

The late 1800s has been described as the low point for local government in our country. Political corruption was rampant. Lord Byron of England wrote at the time, "There is no denying that the government of cities is one of the conspicuous failures of the United States."[19] Against this backdrop, Dillon's court heard arguments that these same governments had powers beyond those granted by the Constitution. Dillon's opinion in the case has been described as "absolutely the most important legal proposition in all of local governmental law."[20]

Simply stated, under Dillon's rule, local governments have only those powers expressly conferred by the state constitution, **state statutes**, or home rule charter; those powers implied in or incidental to the powers expressly granted; and those powers essential to the declared purposes of the municipality.[21]

Code Adoption

We will assume then that the laws of your state or your city charter have enabled the governing body to adopt regulations to protect the public from the threat of fire and explosions. Several states have adopted fire prevention regulations at the state level, leaving local government with only the decision of which agency will enforce them. Either way, in most of the United States, one of the three model fire prevention codes is adopted by the governing body. Hopefully, it is a companion document to the building code in use within the jurisdiction.

The adoption process is structured and is spelled out within the law. Generally, public notice of the jurisdiction's intent to adopt or amend the code is required. Publication of the time and date of public hearings and the availability of the proposed code text for public inspection prior to the hearings are required to appear in a newspaper serving the jurisdiction. The governing body acts on the proposed ordinance after hearing testimony from government officials and interested members of the public. The adopting ordinance also includes an effective date and designates the particular agency or agencies charged with the enforcement of the ordinance.

Figure 1-3. Codes are adopted after a duly advertise public hearing

Two methods of adoption are by **reference** and by **transcription**. Adoption by reference is the more common method, with the adopting ordinance referring to the specific edition of the model code to be enforced. Adoption by transcription requires the entire code to be published within the ordinances of the municipality. The model code groups make their codes available to jurisdictions free of charge for adoption by reference only. Publication of the documents as is necessary in adoption by transcription is only permitted under royalty-bearing licenses granted by the model code groups.

NOTE

The model code groups make their codes available to jurisdictions free of charge for adoption by reference only.

Minimum Codes and Mini/Maxi Codes

In some states, local governments are only empowered to adopt regulations that exceed the requirements of minimum standards adopted at the state level. This system gives local government limited powers.

Under the **mini/maxi** concept, local governments are precluded from adopting any regulations. Uniform regulations are adopted at the state level and cannot be locally amended. Mini/maxi codes are generally favored by architects, builders, and contractors as being good for business by providing uniform regulations. Lobbying efforts can be targeted at the state capital and not spread across the numerous political subdivisions. Mini/maxi codes are generally opposed by fire officials and local governing bodies as an unnecessary intrusion into a local issue by a governing body that may be out of touch with the needs of the people.

NOTE

Under the mini/maxi concept, local governments are precluded from adopting any regulations; uniform regulations are adopted at the state level and cannot be locally amended.

Kentucky, Minnesota, Montana, New Jersey, North Carolina, Oregon, and Virginia are among the states with some form of mini/maxi code.[22] More will follow with our national push to reduce trade barriers. There are positive aspects to mini/maxi codes. It forces the fire service to get our heads out of the sand and participate in the code development process at the state and national levels. It has also been a powerful organizing factor causing departments to organize state-wide training which leads to statewide networking and the sharing of knowledge.

Local Amendment of the Model Codes

Model codes are, by their very nature, designed to address the fire problems faced by most jurisdictions. This is one of the relative strengths of the entire process. Most of the problems in most jurisdictions are addressed by provisions based on our national and regional fire experience. What about the condition then that exists solely in one jurisdiction? Or perhaps the desire of a community to have standards of safety that far-exceed the national norm?

NOTE

Model codes are, by their very nature, designed to address the fire problems faced by most jurisdictions.

These needs are addressed through amendments to the model codes that are made at the state or local level and enacted within the code adoption process. A city or county basically passes an ordinance stating "we hereby adopt the twenty-something edition of this model code with the following changes."

Many jurisdictions have adopted code requirements for sprinkler protection, fire department water supply, emergency vehicle access, and explosives or hazardous materials handling and storage that far exceed the model code provisions. Local amendments must be nondiscriminatory, technically valid, and justified by real need. Regardless of how small any code provision may seem, there is a cost associated in some way. In the end, somebody is going to have to pay the bill.

Summary

The model code process has evolved considerably in its first 100 years in the United States. What began as a method of reducing conflagration potential and insuring the profits of the insurance industry has developed into a system of balancing property protection, life safety, and cost effectiveness.

The three legacy model code organizations, BOCA, ICBO and SBCCI, cooperatively developed the International Codes and then consolidated their organizations, existing today as the International Code Council.

The National Fire Protection Association (NFPA) publishes over 300 codes, standards, and recommended practices, including NFPA 101, *Life Safety Code* and NFPA 70, *National Electrical Code*.

Model codes are developed through consensus processes and are available for adoption by state and local governments. Codes must be adopted in order to be enforced.

Review Questions

1. Name the three legacy code organizations.

    a. _____

    b. _____

    c. _____

2. Before a model code can be enforced within a jurisdiction, it must be _____ by the governing body.

3. Changes to the model building and fire prevention codes are voted on by _____.

4. The fundamental power of the state to place restraints on personal freedoms and property rights is called _____.

5. Name two methods of code adoption.

   a. _____

   b. _____

6. A uniform code that is adopted at the state level and that cannot be amended by local governments is called a _____ code.

7. Which amendment to the United States Constitution is the basis for the promulgation and enforcement of building and fire prevention regulations by state and local governments? _____

8. What umbrella organization was organized by the model code groups to facilitate the development of the international building and fire prevention codes? _____

9. What legal principle states that local governments have only those powers conferred by charter or state statute? _____

10. Model codes are available for local governments to adopt by _____ free of charge.

Discussion Questions

1. What organization developed the first *National Building Code* and why was the development significant?

2. What was a fundamental difference between the process used by the National Board of Fire Underwriters (NFBU) in the development of their codes and that used by the model code groups today?

3. Why are mini/maxi codes favored by many business interests?

KEY TERMS

code

a systematically arranged body of law or rules; when and where to do or not to do something

model code

a code generally developed through the consensus process through the use of technical committees developed by an organization for adoption by governments, e.g., the *International Fire Code, Uniform Fire Code* or the *International Building Code*

adopt

to formally accept and put into effect

legacy codes

The codes developed by the former regional organizations, BOCA, ICBO and SBCCI

ordinance

a law of an authorized subdivision of a state, such as a county, city, or town

standard

a rule for measuring or a model to be followed; how to do something, what materials to use (also known as referenced standard)

Dillon's rule

legal ruling issued by Chief Justice John Forrest Dillon of the Iowa Supreme Court in the late 1800s whereby local governments possess only those powers expressly granted by charter or statute

statute

a law enacted by a state or the federal legislature

adoption by reference

method of code adoption in which the specific edition of a model code is referred to within the adopting ordinance

adoption by transcription

method of code adoption in which the entire text of the code is published within the adopting ordinance

mini/maxi code

a code developed and adopted at the state level for either mandatory or optional enforcement by local governments, and which cannot be amended by the local governments

Chapter 2

Inspection

Learning Objectives

Upon completion of this chapter, you should be able to:

- Describe the limits of the right of entry provisions contained in the model fire prevention codes.
- Explain why warrants must be secured in order to inspect certain buildings and structures.
- Describe the permit and inspection models and the benefits of each.
- Describe the legal requirements for written notices and orders and how freedom of information laws impact fire prevention bureaus.
- Describe how the appeals process within the model codes is established and designed to operate.

INTRODUCTION

The heart of any fire hazard law is INSPECTION, the value of which is generally underestimated as it is not spectacular or colorful. [1]

(Virginia Advisory Legislative Council to the Governor, November 18, 1947)

Just how effective is the inspections process at preventing the occurrence of fire or of minimizing their impact? Some would venture that comparison of any process or premises to "nationally recognized good practice," must result in the identification of deficiencies. If corrections, repairs, or adjustments are made, things simply must be safer. That is probably true, but a more scientific approach was used in a study in the late 1970s that clearly identified the potential value of the inspection process.

In 1978, the National Fire Protection Association (NFPA) published the results of a federally funded study under the title *Fire Code Inspections and Fire Prevention: What Methods Lead to Success?* The study was conducted cooperatively by NFPA and the Urban Institute, with funding and support from the U.S. Fire Administration, the National Science Foundation, and metropolitan fire departments from across the United States. The study was undertaken to determine whether some fire code inspection practices actually resulted in fewer fires, lower fire loss, and fewer civilian casualties. The study focused on all properties covered by fire codes, excluding one- and two-family dwellings.

The study used fire data from seventeen cities and one metropolitan county, and determined that fires due to visible hazards that can be directly remedied by inspectors were responsible for 4–8% of all fires. Fires due to carelessness or foolish actions or electrical or mechanical failures (sometimes due to lack of maintenance) accounted for 40–60% of all fires. The remaining fires (32–66%) were identified as incendiary, suspicious, or natural cause fires, and were considered unpreventable by inspection.[2] When this information was compared to inspection practices, the findings underscored the effectiveness of a routine fire inspection program.

Fire rates appeared to be significantly lower in cities that annually inspected all, or nearly all, public buildings. Jurisdictions that did not annually inspect most public buildings had rates of fires exceeding $5,000, which were more than twice that of those that did inspect them.

Cities that used fire-suppression companies for a large share of their routine fire code inspections had substantially lower rates of fire, presumably because cities that exclusively used fire prevention bureau inspectors did not have sufficient personnel within the bureau to make annual inspections at all public buildings.

Fires due to carelessness or foolish actions or electrical or mechanical failures (40–60% of all fires) were not preventable through direct action by inspectors, because these causes are not readily visible during an inspection. Yet the rate at which these fires occurred was significantly lower in cities with regular inspection programs, indicating that the department that inspects more frequently has more opportunities to motivate occupants. This statistic also underscores the value of company inspections, recognizing the fact that although company inspections cannot be as in-depth, due to the differences in training between inspectors in the bureau and those with other responsibilities, inspections by in-service companies have a significant impact on the incidence of fire within the jurisdiction.

The opportunity to preplan hazards, provide building familiarization, and the chance to establish cordial relations with the community can literally make in-service inspections the excuse that every chief wants to advertise the department. Training for company inspections should be prefaced with the fact that the goals of in-service inspections programs are building familiarization to increase firefighter safety, fire and hazard reduction, and community relations—in that order.

The inspection process is the very backbone of the fire prevention program. Having a fire prevention code in force within a jurisdiction is of little benefit to the public without an effective inspection and enforcement program. Unlike the Vigiles of the first century in Rome, the fire inspector's duties do not include punishing those guilty of violating the code or causing a fire through negligence. The inspector's duties and responsibilities in code enforcement are outlined within the body of the code.

NOTE

The inspector's duties and responsibilities in code enforcement are outlined within the body of the code.

How the inspector performs those functions and their or her conduct in the performance of those duties are governed by various federal, state, and local laws. From local ordinances that prohibit conflict of interest for public officials, to state tort claims laws, to federal civil rights statutes, fire inspections personnel are subject to myriad laws in the day-to-day performance of their duties. Up-to-date information regarding legal developments within the code enforcement field, periodic training in the subject, and competent legal advice are absolute necessities for the fire code official. This text touches lightly on the legal aspects of code enforcement, concentrating instead on the technical aspects of fire code inspection and enforcement. The need for training in the legal aspects of code enforcement cannot be overemphasized.

NOTE

Up-to-date information regarding legal developments within the code enforcement field, periodic training in the subject, and competent legal advice are absolute necessities for the fire code official.

LEGAL ASPECTS

Neither the *International Fire Code* nor the *Uniform Fire Code* specifically mandate the inspection of specific occupancies. They *authorize* the fire official to inspect buildings and premises in order to enforce the provisions of the code. Neither code contains a **duty to inspect clause** stating that all or particular classes of occupancies must be inspected. Two of the legacy fire codes contained language that seemed to imply that there was a duty to inspect:

> *The code official shall inspect all structures and premises, except single-family dwellings and dwelling units in two-family and multi-family dwellings for the purposes of ascertaining and causing to be corrected any conditions liable to cause fire, contribute to the spread of fire, interfere with fire-fighting operations, endanger life or any violations of the provisions or intent of this code or any other ordinance affecting fire safety. (F-108.1, 1999 BOCA National Fire Prevention Code)*

> *The fire prevention bureau shall inspect, as often as necessary, buildings and premises, including such other hazards or appliances designated by the chief for the purpose of ascertaining and causing to be corrected any conditions which would reasonably tend to cause fire or contribute to its spread, or any violation of the purpose or provisions of this code and any other law or standard affecting fire safety. (103.3.1.1, 2000 Uniform Fire Code)*

These code sections clarified the intention of code writers that fire codes were applicable to all structures. They were not found by courts to impose liability on jurisdictions for failure to inspect any or all buildings.

Under common law, local governments have not been held liable for injuries resulting from enforcement or failure to enforce statutes or ordinances, as was clearly underscored in *Grogan v. Commonwealth*, a suit arising out of the disastrous Beverly Hills Supper Club fire in 1977. The City of Southgate, Kentucky, and the State Fire Marshal's Office were named as defendants arising from their failure to enforce the state building code and that the failure resulted in 165 deaths. Both the city and state were dismissed from the suit during preliminary motions by the trial court. The Kentucky Supreme Court later upheld the dismissal.[3]

Applicability of the Code

Model fire codes are intended to apply to all buildings, structures, and premises:

Uniform Fire Code

> *This Code shall apply to both new and existing conditions, (2006 UFC 1.3.1) . . . permanent and temporary buildings, processes, equipment, systems, and other fire and related life safety situations (2006 UFC 1.1.1).*

International Fire Code

> *Conditions and operations arising after the adoption of this code, and Existing conditions and operations. (2006 IFC 102.2)*

A common misconception is that single-family dwellings and the dwelling units in multiple-family dwellings are exempt from the provisions of the code. This misconception is a serious misinterpretation of the code or of the Fourth Amendment that could lead to dire safety consequences for the public as well as potential liability for the fire official.

> **NOTE**
>
> **A common misconception is that single-family dwellings and the dwelling units in multiple-family dwellings are exempt from the provisions of the code.**

The model codes do not exclude dwellings or dwelling units from their scopes. Dwelling units are only exempted from routine inspection. Timothy Callahan, writing in *Fire Service and the Law*, describes the balance between the legitimate police power of the state and the public's right of privacy:

> Subject to issuance of appropriate notice, there is no doubt that the right to inspect a private dwelling could be considered a reasonable exercise of the police power. *This is true where there is reason to believe that conditions exist which make the property a fire menace to the community.* But where a residence is so isolated from other occupancies or properties that it would not be considered a hazard, even if on fire, the right of privacy of the dwelling's owners might be considered paramount to the police power.[4]

However, Supreme Court decisions such as *Mapp v. Ohio*, *Frank v. Maryland*, and *Camara v. Municipal Court* have clearly underscored the limits on police power imposed by the Fourth Amendment. The nonconsensual warrantless entry of a dwelling is an especially serious breach, unless the entry is in response to an emergency.

NOTE

The nonconsensual warrantless entry of a building is an especially serious breach, unless the entry is in response to an emergency.

In *Michigan v. Tyler*, the Court held that a "burning building clearly presents an emergency of sufficient proportions to render warrantless entry reasonable. Indeed, it would defy reason to suppose that firemen must secure a **warrant** or consent before entering a structure to put out the blaze."[5]

The protection of the Fourth Amendment is not limited to dwellings. In *See v. City of Seattle*, the Court held that "the businessman, like the occupant of a residence, has the constitutional right to go about his business free from unreasonable entries upon his private commercial property."[6] The right of the businessman to deny government officials access to those areas of his premises that are not open to the public is protected by the Fourth Amendment. Unless permission to enter and conduct an inspection is obtained, a warrant must be secured by the inspector.

Right of Entry

Each of the model codes contains a right of entry clause that permits the fire official to enter all structures and premises at reasonable times to conduct inspections. The inspector's right of entry is limited however:

- The fire code cannot authorize an unconstitutional inspection. The inspector must identify himself, present credentials, and seek permission to conduct the inspection. Permission to inspect must be obtained or an inspection warrant must be secured prior to conducting the inspection, unless the inspector has reason to believe that a life-threatening condition exists.
- Inspections must be aimed at securing or determining compliance with the code, not with gathering information for other purposes such as the enforcement of other laws or ordinances.[7]
- Permission to conduct inspections must be requested during reasonable hours, normally the hours that the business is open or staff is present.
- Inspection of those areas visible from the public way does not require permission from the owner.

NOTE

The protection of the Fourth Amendment is not limited to dwellings.

INSPECTION PRIORITIES AND FREQUENCY

Although the model codes mandate the inspection of most structures and premises, they do not establish inspection priorities or frequency. These issues must be determined by the local jurisdiction based on need and available resources.

Inspecting every premises and occupancy on a regular basis is fine if you are employed by a small enough jurisdiction, or one that has the staffing to carry out that many quality inspections. But how often is enough? Who needs to be inspected first? Why inspect some occupancy types and not others? These important policy issues must be addressed by the senior managers of the fire prevention bureau.

Two models used in determining inspection priorities are the permit model and the inspection model.[8] Regardless of the system used, it should target potential life safety and property hazards first. It should be fair in that all occupancies within a given classification or type, are inspected with the same frequency. Administrative procedures that outline the process should be a written part of the bureau's standard operating procedure.

The Permit Model

Perhaps the simplest model to get started and the easiest to defend as technically valid is the permit model, using the permit requirements and thresholds directly out of the fire prevention code.

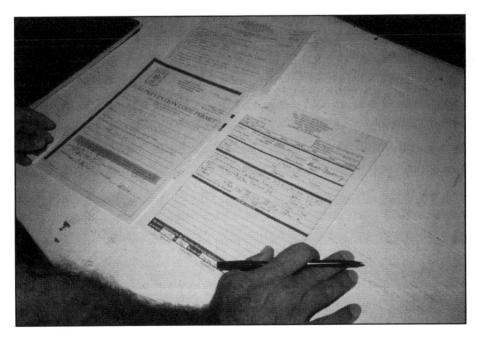

Figure 2-1. The Permit Model is the easiest to defend in court.

The list of required permits includes occupancies with a high potential for life loss and those involving hazardous processes or the storage and handling of hazardous materials. Like all provisions of the model codes, they are the result of our nation's fire experience and the collective wisdom of the membership of the model code group. How can you argue against that?

Permits are required to maintain nightclubs because there were fires like the Coconut Grove, Station and the Beverly Hills Supper Club. Buildings housing processes that employ flammable liquids and gases and storage occupancies that handle aerosols and other hazardous substances have all been the site of fires that resulted in the loss of life, extensive property damage, and often the catastrophic disruption of the entire community. They all require permits to operate. Those occupancies that have traditionally enjoyed a low incidence of fire, such as business and professional offices, banks, and small mercantile occupancies do not require permits to operate.

NOTE

The list of required permits includes occupancies with a high potential for life loss and those involving hazardous processes or the storage and handling of hazardous materials.

The permit process prioritizes inspections for the code official and enables the fire official to identify those occupancies in which fire operations personnel are most likely to encounter hazardous conditions or life hazards. This information should be shared with operations staff. Occupancies requiring permits are those that, in the opinion of the model code groups, have the most potential for fire and life loss.

The burden of application for permit is on the business owner. It is his or her obligation to apply for a permit prior to commencing operations. It is not the responsibility of the fire official to find the business owner. Failure of the business owner to permit inspection at a reasonable hour is grounds for permit revocation.

NOTE

Permits are required to maintain nightclubs because there were fires like the Station, Coconut Grove and the Beverly Hills Supper Club.

The permit itself can be a valuable tool for gaining access to structures and premises. Permit fees can be used to help recover the cost of providing inspection services.

The Inspection Model

In the inspection model, the fire prevention bureau determines which types of occupancies will be inspected, identifies them, and conducts the inspections. Written procedures are needed to guard against claims of harassment and selective enforcement by some business owners. If a determination is made to inspect public assemblies during the month of June, then all public assemblies should be inspected throughout the jurisdiction.

Deviation from the schedule is only justified by legitimate circumstances. The failure of several range hood suppression systems in restaurants might warrant special inspections of restaurants outside of, and in addition to, the normal schedule. These can easily be justified based on the actual fire experience of the jurisdiction.

Advantages of the inspection model process include:

- It enables the bureau to pick and choose those occupancy classifications to be inspected.
- It is less labor intensive than the permit process, requiring fewer administrative functions.
- There is more flexibility in the quality of personnel assigned to perform inspections.[9]

When using the permit system, inspectors must have the ability to determine compliance with the entire code. Under the inspection model, only one type of occupancy is inspected at a time, theoretically limiting the amount of the code to be applied.

NOTE

Using the inspection model, an uneducated or untrained inspector essentially is not noticed until a disaster occurs.

In *Political and Legal Foundations for Fire Protection*, Howard Markman and coauthors caution code officials about the inspection model, where "using the inspection model, an uneducated or untrained inspector essentially is not noticed until a disaster occurs."[10]

Supervisory personnel can be lulled into thinking that all is well because they are not getting requests for help from the inspector and the telephone is not ringing with calls from business owners. The use of untrained or inadequately trained inspections personnel is a recipe for trouble. Better to do less-inspections, than less-than-adequate inspections. The jurisdiction's liability in not performing an inspection at all is questionable unless the statutory duty to perform an inspection is explicit.[11] The potential liability of the jurisdiction, the fire official, and the inspector for a negligent inspection is another matter.

NOTE

Better to do less-inspections, than less-than-adequate inspections.

INSPECTION WARRANTS

The Supreme Court has held that warrantless entry by government officials can constitute a serious breach of the Fourth Amendment rights of the occupant of a building. In *Camara v. Municipal Court of San Francisco*, the occupant of an apartment refused to allow housing inspectors to make a routine inspection, as required by city code. He was later arrested for refusing to permit a lawful inspection. In overturning the lower courts, the Supreme Court held that a search of private property is unreasonable unless consented to or authorized by a warrant.[12]

The Court did, however, acknowledge that routine inspection of the physical condition of a property is a less hostile intrusion than a typical policeman's search for the fruits and instrumentalities of a crime.[13] As a result of *Camara* as well as other Fourth Amendment cases, many state legislatures have adopted statutes that establish procedures for obtaining administrative or inspection warrants.

In order to secure an administrative warrant, the inspector must convince an impartial magistrate that either the premises should be inspected based on reasonable legislative or administrative standards or the inspector has reason to believe that a particular violation of the code exists at the premises.

The written administrative procedures that were discussed in connection with the permit and inspections models are critical elements in securing an inspection warrant. There are only two acceptable answers when the magistrate asks why he or she should issue a warrant to inspect a premises. Either you know of a specific violation of the code and require access to conduct a thorough inspection or, based on the written procedures that you can cheerfully provide, the occupancy is due to be inspected.

It is important to remember that the magistrate represents the people and is supposed to be a neutral party. They are not supposed to "be on your side." Don't take it personally if they make you prove your case a little more than you think is necessary. It is the magistrate's job to ensure that your request is based on a legitimate government interest. Do not kid yourself into thinking that selective code enforcement, in which certain occupancies or owners are targeted for enforcement efforts for reasons that are not completely legitimate, is not attempted in different communities from time to time.

NOTE

It is important to remember that the magistrate represents the people and is supposed to be a neutral party.

CONDUCTING INSPECTIONS

Most inspections are conducted with a representative of the owner present and should generally be conducted during the business's nonpeak hours (see Figure 2–2). However, public assembly occupancies such as nightclubs, theaters, and exhibit halls also must be inspected during hours of peak occupancy to ensure no overcrowding exists, the exits are clear, and fire access lanes are clear. If the facility is industrial or occupied by the federal government, you may also be accompanied by safety or union representatives, or both. At times, an owner's representative, attorney, or even a reporter could be on the scene. Remember, you are conducting the inspection. You are setting the pace and determining the direction to be followed, and you should be asking the questions. Do not allow the others, no matter how well-intentioned, to distract you. Do not allow the inspection to turn into a sideshow.

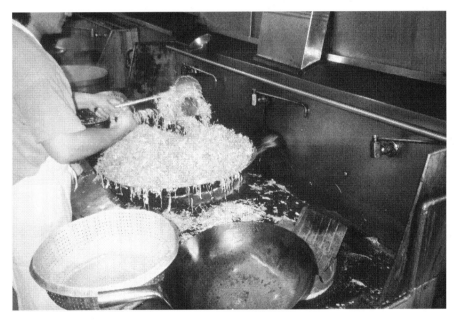

NOTE

A fair word of advice to almost any inspector is to slow down while conducting inspections.

Figure 2-2. Routine inspections of restaurants should be conducted early in the day or during slow periods, not at peak times.

If you find yourself losing control of the inspection, call the thing off. Come back later, after you have had time to confer with your supervisor, legal department, or just had time to get your act together. Time is on your side, and as you will see throughout this text, you've got all the cards you need to play the game. A fair word of advice to almost any inspector is to slow down while conducting inspections. Talk less, listen more. Ask more questions. Stop talking and think about what you are seeing. *You can't hear with your mouth open.*

Ideally, inspections should begin with a review of the records of previous inspections of the facility. That may not be possible, however, especially if responding to a complaint received in the field. A thorough inspection of the outside of the building including fire department access, hydrant accessibility and condition, outside storage conditions, and the viability of the exit discharges should always be attempted before you even get out of the car. Drive the long way around the parking lot, or better yet, once around the block.

Take your time and look at the place. Are trucks blocking the fire department connection? Do parked cars block exit doors? How does the outdoor storage of idle pallets and other combustible material affect exiting and building exposures? Complete your inspection of the outside of the facility on foot, after entering the building and meeting with the owner or manager.

Your inspection of the inside of the facility should include every room and space within the building. This condition should be made clear at the beginning of the inspection so that the owner or manager brings all appropriate keys along on the inspection. Pick your route through the building at the beginning of the inspection and remember the layout of the structure that you noticed while outside. Many an inspector has been led on a circuitous tour by a building owner and shown only what the owner wanted the inspector to see.

NOTE

Your inspection of the inside of a facility should include every room and space within the building.

It is better to ride the elevators up and walk down the steps. Alternating stairwells between floors saves some time. An important point to remember is that the area you might skip because it is inconvenient or difficult to access, is probably the same area that the last two inspectors skipped for the same reason. Although I am not advocating a full-blown top to bottom, roof to basement inspection of every structure every time, such an inspection certainly should be undertaken at given intervals. This policy issue must be determined on a management level, based on staffing resources, workload, and fire experience within the jurisdiction.

NOTE

The area you might skip because it is inconvenient or difficult to access is probably the same area that the last two inspectors skipped for the same reason.

The general condition of the premises also dictates how in-depth the inspection must be. A facility that evidences numerous hazardous conditions or illegal construction must be carefully and thoroughly inspected, including building voids, utility shafts, plenums, and other spaces that might not be carefully examined on every inspection.

Every inspection should include the following minimum components:

- Inspection of the outside of the structure for fire department access, hydrant and fire department connection access and condition, outside storage, and exit discharge viability.
- Inspection of the building fire protection systems to determine they have power, all valves are open and appear to be in general good order, and have current maintenance records. This part includes a trip to the sprinkler room to determine that heat is provided for the sprinkler riser and a trip to the fire pump to verify that the pump is being maintained and run.
- Inspection of the means of egress in the structure, including all interior and exterior exit stairs. Every exit door should be opened and all door hardware should be examined for function and appropriate application. All emergency means of egress lighting should be inspected and tested.
- Inspection of all hazardous processes, hazardous storage, or other special hazards. This part of the inspection is occupancy-specific and can range from chemical hazards in industry to the use of decorative materials in public assemblies.
- Inspection of all mechanical, electrical, and machine rooms.
- Inspection of rated assemblies for penetrations and continuity and detection of construction that appears to be inconsistent with the building construction type.

REPORTING INSPECTIONS

Most fire prevention bureaus have inspection report forms that comply with legal requirements of the jurisdiction. They may advise the building owner of his right to appeal, refer to a section of the municipal code that authorizes the inspection, or even mention the penalties for noncompliance. Philosophies on the subject range the entire spectrum, from no-nonsense heavy-handed to customer friendly. Some are a combination of Report of Inspection/Notice of Violation.

Some jurisdictions develop checklist inspection forms. They can be very useful, especially for complex processes and for fire protection system installations. A note of caution is in order, however. Checklists should never be considered substitutes for the actual code documents! The model codes and referenced standards are expensive. Over the years, some organizations have attempted to reduce the cost of outfitting inspectors by substituting lists and checklists of code requirements. This is as risky as it is foolhardy. The code documents are the tools of the trade for an inspector. A tradesman who has to borrow tools or who uses less costly and inferior substitutes can never be expected to produce quality work.

The checklist is basically a correlation of code requirements; a user-friendly tool to help the inspector verify specific items regulated by the model codes and referenced standards. Use checklists to guide your inspection and to aid in coordinating the requirements of different standards that may apply to a particular process or system. Be sure to refer to the actual code or standard before issuing a notice of violation, particularly if it involves the court system.

Code officials who caution against using checklists usually list objections such as the checklist becomes a crutch that prevents inspectors from developing proficiency in using the code documents. Checklists could be improperly developed and reference incorrect editions of the various codes and standards, and the inspectors have a tendency to check only the items listed on the form and nothing else.

NOTE

Every inspection and every visit should be documented.

Officials who endorse the proper use of checklists reply that they are simply additional tools for the well-trained and well-equipped inspector. Both sides cite valid concerns. These issues are as old as the trade itself. There has never been a substitute for good training and proper equipment in any trade or craft.

Every inspection and every visit to a building should be documented, including visits where access is denied or the inspection is rescheduled at the request of the owner. These documents may be critical elements in obtaining an inspection warrant in the future. Reports must be legible and clearly state violations noted and corrective action required. By merely looking at your report, will the judge understand what you were ordering the building owner to do? If your correction order is not clear and concise enough to pass this test, you could be 100 percent right and walk out of court—as the loser.

NOTE

Reports must be legible and clearly state violations noted and corrective action required.

Should your report end up in the hands of a jury, they are going to judge your competence based on what they know. They know very little about fire prevention and protection, but they do know basic grammar, spelling, and punctuation. For better or for worse your report says as much about you as your appearance and demeanor.

VIOLATIONS

Notices of Violation

Written notice of violation to the owner or occupant of a building or premises was required by the legacy fire codes and still mandated by the *Uniform Fire Code*:

> *Whenever the AHJ determines violations of this Code, a written notice shall be issued to confirm such findings. (2006 UFC 1.16.1)*

> *The International Fire Code merely authorizes the fire official to issue a notice:*

> *When the fire official finds a building, premises, vehicle, storage facility or outdoor area that is in violation of this code, the fire official is authorized to prepare a written notice of violation . . . (2006 IFC 109.2)*

The use of the term shall indicates that these are mandatory actions or requirements under the law. If a law or code states that something **shall** be done it is a "positive and definitive requirement. If the word **may** is used, discretion is allowed."[14] Many jurisdictions rewrite the administrative chapter of the model fire codes to ensure that it complies with their state statutes. Inspectors should have a clear understanding of their authority, duties, and responsibilities prior to conducting the first inspection.

> **NOTE**
>
> **The use of the term shall indicates mandatory actions or requirements under the law.**

In 1970, five persons perished in a fire at the Gold Rush Hotel in Anchorage, Alaska, eight months after the state fire marshal's office had inspected the hotel and detected numerous code violations. The resulting lawsuit alleged the state fire marshal's office had promised, but failed to provide the owner a letter with the required corrective actions needed. The Alaska Supreme Court upheld a lower court ruling of **negligence**, in that the fire official's duty to provide the written corrective order in this case was not discretionary, but mandatory.[15]

> **NOTE**
>
> **If the word may is used, discretion is allowed.**

The model codes also include specific methods of delivery, which include personal delivery, delivery by mail to the last known address, and posting the notice to the structure. Care must be taken to ensure that all appropriate requirements are complied with. Personal delivery is always the most desirable method, followed by certified mail. Posting the structure is least desirable and invariably leads to claims in court by the owner that he "certainly would have complied if he had only known."

The code provision that authorizes posting the notice on the premises is actually there to protect the code official. Since the code states a notice *shall* be served, the code official's inability to track down the owner could potentially lead to liability. Hence, if all else fails, posting the notice at the entrance at least allows the official to perform a required act.

Elements of a Notice of Violation

Notices of violation that follow a prescribed format, such as a fill-in-the-blank printed form are preferable to a letter. The form can be preapproved and used by the entire bureau. The form must comply with the requirements of the code. This may include a statement regarding the right to appeal. As a minimum, notices of violation should:

- Be in writing and include the date, legal address of the premises inspected, and name of the inspector.
- State the specific violations noted by the inspector and identify the applicable code section for each violation.
- State the required corrective action.
- List the date on which a follow-up inspection will be performed to ensure compliance.

NOTE

Personal delivery of notices of violation is always the most desirable method, followed by certified mail.

GAINING COMPLIANCE

Enforcement actions for noncompliance depend on your state law, the local ordinances of your jurisdiction, and the policies of your local government and department.

Criminal Charges

In some jurisdictions, violations are criminal misdemeanors and inspectors are empowered to issue **summonses** or **citations** for violations of the fire code. This method is effective in that it quickly gets the owner's attention. There is a downside however. You are bringing criminal charges against an individual; this is not traffic court. Most prosecuting attorneys have little knowledge of fire or building codes and even less time or interest in learning about them. If you want to win your case, you need to do your homework and provide the prosecutor with a succinct, one-page script or summary that describes conditions and the potential danger in plain English. Include the sections of your state, county, or municipal code that adopt the model code, the sections that empower you to enforce it, and the specific section of the fire code that has been violated. We'll discuss courtroom demeanor later in this chapter.

Civil Charges

An **injunction**, or order from a court with jurisdiction prohibiting a specific action (in this case a code violation), can be effective in gaining compliance. If the violation continues, the violator is in contempt of court and can be jailed until compliance is attained.

The downside is that the preparation of a motion for injunctive relief must be prepared by your jurisdiction's legal department, and a lawyer must present the case in court. This is an extremely drawn out, labor-intensive process. Occasionally, a case can become so political or so complicated that injunctive relief is the remedy of choice.

In some jurisdictions, civil penalties can be assessed for code violations, often without taking the violator to court. Civil fines often act as a deterrent, but to some, they are merely the "cost of doing business."

APPEALS

The appeals process is a two-edged sword. Use it wisely and it can aid in your code enforcement efforts and keep the code official out of a lot of hot water. Abuse it or fail to respect the potential damage it can cause and it will become a nightmare come true. The elements of the process, including who may file an appeal, the application process, and the time limits for each step are clearly prescribed by the codes or within adopting ordinances. Failure to comply with any of the provisions are grounds for dismissal of the appeal, if the appellant has violated the process; or a finding against the code official if he has violated the process. At the very least, the local official who cannot follow his own process gives the jurisdiction a black eye.

> **NOTE**
>
> **Failure to comply with any of the provisions of proceeding with an appeal are possible grounds for the dismissal of the appeal.**

Appeals Boards

Each of the model fire codes has different provisions for the makeup of the boards. The IFC establishes the board in the administrative provisions of Chapter 1. Additional information on appeals boards is included in Appendix A of the IFC. Appendices are generally informational and are not adopted as part of the code. The UFC gives the jurisdiction more flexibility in determining the makeup of the board, but not more than one of the members may be engaged in the same business or profession.

Some jurisdictions include members of the real estate industry, attorneys, and citizens at large. Often such appointees lack the technical expertise in the code to render competent decisions and fall back on their political or social beliefs. This is unfortunate and should be avoided.

NOTE

Boards of appeals are responsible for hearing appeals on the technical aspects of the code.

In some jurisdictions the same board may hear appeals from both the building and fire code official's decisions. It is very important that both the makeup of the board and the scope of their responsibility is clearly stated by the local governing body when the board is constituted. Boards of appeals are responsible for hearing appeals dealing with the technical aspects of the code, not with the political or philosophical movements that may have led to code development.

Appeals boards cannot and should not debate the validity of a portion of the code. That is for the local governing body or the courts to decide. The zealot, with a political ax to grind, has no place on an appeals board. The citizen whose personal conviction that less government regulation is better or that the government needs to rectify some past social wrong and casts his vote based on political belief, is a danger to public safety. Such behavior undermines the code process.

Appeals Hearings

Hearings are public meetings that must comply with state and local laws that regulate government business. Hearings are generally advertised in newspapers or government agendas with public distribution and held in public buildings. The proceedings are recorded or transcribed and become public record.

Although hearings are structured and run by the chairman, they are generally informal enough to ensure that all parties have an adequate opportunity to present their case. The chairman is responsible for ensuring that the hearing remains on track and does not become a session of bullying the public officials involved. Once arguments have been made and rebutted by both sides, members of the board pose questions to the appellant and code official before deliberating and rendering a decision. Appeals of the decision may go to a state board or to the courts if either party is dissatisfied or feels the system was not followed.

Hearing Preparation

"It is this way because we're the fire department and we say so," won't get you very far before the appeals board. The fire department "mom and apple pie" is a great angle though, and you should make the most of it, but you must win the appeal on the merits of the issue.

Cooperation between building and fire officials is critical when it comes to the appeals process. When the building official testifies that an illegal locking arrangement is unsafe because it is a code violation, the board gets one message. If the next witness is a fire officer in dress uniform who testifies that the locking arrangement is unsafe because it precludes the use of required exit doors in the event of an emergency, potentially hampers fire department operational efforts, and potentially endangers the building occupants, your one-two punch will be very effective.

NOTE

Cooperation between building and fire officials is critical when it comes to the appeals process.

When appearing before the board:

- Provide board members all the documents involved in the appeal as well as a brief one-page synopsis of the issues.
- Have adequate, well-prepared handout material for all board members as well as the appellant and members of the public that may be present.
- Your dress, speech, and mannerisms are crucial elements of your presentation at the hearing. It may be advantageous to have several witnesses for the jurisdiction, such as the fire marshal in civilian dress, a fire officer who will speak to operational issues in dress uniform, or even a fire investigator. The board probably will not understand your department's organizational structure, so use it to your advantage.
- Be cordial and polite to the members of the board, the appellant and his attorney or representatives, and any witnesses present. A handshake and hello before the hearing and at the conclusion will help set the tone for non-hostile hearings that stick to the technical issues.
- Do not take the process or the results personally. Sometimes logic seemingly goes out the window at such hearings. Critique your presentation and make the next one better.

Courtroom Preparation and Demeanor

Preparation for court isn't much different than preparing for the appeals board. Court proceedings are much more formal, the pace is often surprising, due to the volume of cases on the docket. It's easy for a new inspector to get rattled.

Police officers, animal control officers and code officials who deal with property maintenance and zoning issues often have busy case loads and assigned court days. If you cannot sit in on fire code cases as a spectator, sit in on several property-maintenance and zoning cases. Watch how the inspectors interact with your jurisdiction's legal counsel or the prosecutor. Don't be afraid to ask the inspectors for advice after court. Pick the brains of the inspectors in your office.

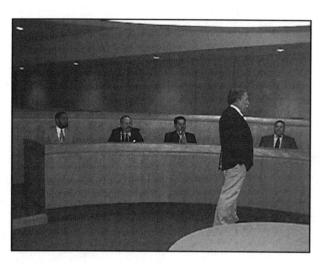

Figure 2-3. Courtroom demeanor is important. Your dress uniform or coat and tie are the uniform of the day. Do not put your hands in your pockets! Courtesy of Duane Perry.

Observe the protocols for the court (Figure 2-3) . Most courts prohibit cell phones; leave yours in the car. Do not chew gum. Do not put your hand in your pockets. Dress smartly. You'll be judged by your looks and your demeanor. Nothing screams unprofessional more than a sloppy, disheveled appearance.

Arrive early and meet with your legal counsel. In most jurisdictions, your first interaction with your counsel will be immediately before court, unless the case is egregious. Have the case boiled down to a single sheet of paper with names, dates and addresses. Have the case boiled down to a few, understandable sentences without technical jargon. Have someone in your department who isn't an inspector read it and ask them to describe the situation. If it's not what you meant, re-write it. Remember, you are writing a script for your legal counsel. Make sure to include:

- Date(s)
- Location
- Proper legal name(s)
- Business name
- Unsafe condition(s) with code section(s)
- Your actions and orders

- Any statements made by the defendant (including insults or threats--verbatim)

It's wise to have the date(s) of your certification(s) with you, if required by your state, as well as the date you were sworn in.

When your case is called, stand and take your place. Speak only when spoken to or asked a question. When questioned by the defense attorney address your answer to the judge or jury. Pause a second to allow the judge to disallow the question or for your attorney to object. This is especially important if the defense attorney goes off track or attempts to discredit you. If you find yourself getting rattled, look to your attorney. Do not allow the defendant's attorney to make it personal. This might be your case, but the issue is not your problem. You're just the inspector who found it and did your duty to have it corrected.

If the defendant prevails, but corrected the conditions before trial, or agrees to comply with the code and make repairs, you really won. You might have been in criminal court, but unlike the shoplifting, loose dogs, and simple assault cases, the object is to gain compliance not exact punishment.

RECORDS

A fire prevention bureau lives and dies by its records. Reports of inspection, notices of violation, code modification letters, and code interpretations are but a few of the documents that must be maintained. Most states have retention schedules for public documents that mandate the storage of certain records for fixed periods of time. Failure to comply is a violation of state statute or local ordinance, as well as an embarrassment. By note following the retention schedule, you may be violating your own code.

Freedom of Information and Public Access Laws

Most states have adopted laws that guarantee public access to the workings of government. Many of these were patterned after the **Freedom of Information Act** signed by President Lyndon Johnson in 1967.[16] The "public" had a lot less to do with the passage of these laws than the press, but the laws have opened the doors on many government functions. Every person who holds a government position, whether compensated or not, should have a clear understanding of these regulations and the impact on their duties.

Generally, all documents in the possession of government are public records, and as such must be made available for public inspection and copying. There are specific exceptions such as personnel records, medical records, ongoing criminal investigations, and others, depending on your state statute. The federal statute permits exemptions in only nine specific areas including national defense, law enforcement files, certain personnel files, and other information whose disclosure is prohibited by statute.[17] A copy of the law and guidelines for compliance are available from some state attorney general's offices. Check with your jurisdiction's legal officer.

NOTE

Generally, all documents in the possession of government are public records.

Requests for information are often subject to very strict time constraints in which government officials must respond within a given number of days or risk court sanctions. A response that the records are not available in the form requested may be adequate, if it is made within the allotted time frame. Freedom of information laws:

- Make documents, recordings, videos, and other media in the possession of government open for public inspection. Exceptions include ongoing criminal investigations, medical and personnel records, and attorney–client communications.
- Require government to respond within a given time frame to requests for information.
- Require that all meetings of boards, commissions, and elected bodies be open to the public and advertised in advance.
- Do not require government agencies to create documents such as studies, comparisons, or summaries in response to requests. Requests are limited to records that exist in their current form.

NOTE

Freedom of information laws contain specific penalties for government officials who fail to comply.

These laws contain specific penalties for government officials who fail to comply, and they can be quite stiff. Review the law in effect in your state. If your agency does not have a written standard operating procedure for requests for information, it is time to get one. The first time you face the judge to explain why a reporter was denied access to an inspection report, you'll wish you had.

Summary

The inspection of buildings and premises is the real backbone of a fire prevention program. The fire code drives the inspections process by establishing the scope of the process, right of entry to inspect, and actions to be taken when noncompliance is discovered.

All structures fall within the scope of the code, although one- and two-family dwellings and the dwelling units of multifamily dwellings are exempt from routine inspection. Administrative or inspection warrants are required to gain access to residential and commercial property when entry is denied.

Two methods of determining inspections priorities are the permit model, in which permits are required for certain hazardous processes and occupancies, and the inspection model, in which certain occupancy types are inspected throughout the jurisdiction at a given time.

Inspections should begin with the outside of the structure or premises. Every inspection should include the building fire protection systems, viability of all portions of the exits, fire protection elements of the structure, building utilities, hazardous processes, and issues affecting the life safety of the occupants. Every inspection should be documented on the proper form.

The preparation and issuance of notices of violation are a mandatory requirement of the model fire prevention codes. Specific provisions in the case of noncompliance vary according to locality. Criminal charges, civil actions, or both are available to some jurisdictions.

Appeals from decisions of the fire official are heard by boards established by the model codes in similar processes. The makeup of the boards as well as the scope of their responsibility and authority are crucial to their effectiveness. Fire officials must carefully prepare for hearings and should coordinate their efforts with the other code officials of the jurisdiction.

Various freedom of information laws have been passed by the states, which guarantee public access to records in the possession of public agencies and require government business to be conducted in the public forum. The laws contain penalties that the court may impose if government officials are found guilty of violating the statutes. Every bureau should have a copy of the statutes in effect within the jurisdiction, as well as written standard operating procedures for handling public requests for information.

Review Questions

1. A _____ clause is a code provision that mandates the inspection of certain occupancies.
2. The right of entry to inspect a premises is limited by the _____ Amendment of the Constitution.
3. When denied entry to perform an inspection, what remedy is available to the fire prevention bureau? _____
4. Name two administrative methods of determining inspection priorities.

a. _____
   b. _____

5. The use of the term _____ indicates that a provision is mandatory, and discretion on the part of the fire official is not allowed.
6. As a minimum, notices of violation should consist of what four elements?

   a. _____
   b. _____
   c. _____
   d. _____

7. When a court with jurisdiction issues an order prohibiting a person from violating the fire prevention code, this order is called a(n) _____.
8. Appeals boards should render decisions based on

   a. political considerations.
   b. economic considerations.
   c. technical merit.
   d. social justice.

9. Name three types of records that are generally exempt from disclosure under state freedom of information laws.

   a. _____
   b. _____
   c. _____

10. Are government agencies obligated to create reports or summaries of information when requested under state freedom of information laws? _____

Discussion Questions

1. Your office has received a complaint from a citizen who claims that her neighbor is storing a 55-gallon drum of racing fuel in the basement of his townhouse. Upon questioning, she accurately describes the red flammable liquid label on the drum. Does the fire prevention code apply in this situation? If denied entry by the homeowner, what options are available to gain entry? What effect could the complainant's desire to remain anonymous have on your ability to gain access to the townhouse?
2. A local fire protection contractor has requested copies of inspection reports for a list of specific occupancies that cancelled service contracts with his company in the preceding year. Is he entitled to these records under the freedom of information law?

KEY TERMS

warrant

a legal writ issued by a judicial officer commanding an officer to arrest a person, seize property, or search a premises

shall

indicates a positive and definitive requirement of the code that must be performed; action is mandatory

may

indicates a discretionary provision of the code; enforcement is left to the judgment of the fire official

negligence

culpable carelessness; the failure to act as a reasonable and prudent person under similar circumstances

summons

a written order, issued by a judicial officer, law enforcement officer, or other authorized official, directing an alleged offender to appear in court at a specific time to answer a criminal charge; a summons issued by a law or code enforcement officer is often classified as a **citation**

citation

a written order, issued by a law enforcement officer or other authorized official, directing an alleged offender to appear in court at a specific time to answer a criminal charge

duty to inspect clause

a clause within a code or ordinance that requires the code official with jurisdiction to inspect certain occupancies

injunction

a legal order issued by a court that commands a person or entity to perform a specific act or prohibits a specific action by that person or entity

Freedom of Information Act

a law passed by Congress and signed by President Lyndon Johnson in 1967 that, as well as the state laws patterned after it, guarantees public access to all documents and information under the control of government, with certain specific exemptions

Section 2

THE BUILDING CODE

The model building codes are lengthy documents. They have to be in order to regulate everything from concrete aggregates and elevator hoist way doors to accessibility for the physically disabled. Do not let that bother you. A fire inspector needs a good working knowledge of about one-third of the building code and the ability to find his way around the rest.

Unless you understand how the egress facilities and fire protection features of buildings are designed and constructed, you cannot possibly ensure that these features are intact and functional during your inspection. Knowledge of building design and construction is crucial for a competent fire inspector. It will also help you, the fire inspector, develop a professional working relationship with the building department. Professional expertise is respected in every profession.

One of the most overlooked aspects of building code knowledge is its value to the fire officer. New construction materials and technologies aimed at reducing costs and conserving natural resources and energy affect fire behavior. Operational considerations at a basement fire in a new, two story single family dwelling aren't the same as at a basement fire in a home constructed in 1970. Today's construction materials are less expensive, lighter, as strong or stronger than the dimension lumber from a generation ago-- as the assembly is intact. They are prone to rapid, often catastrophic failure when compromised by the fire. The notion that firefighting and fire incident management can somehow divorce itself from the construction codes that dictate every aspect of the environment firefighters ply their craft is ludicrous.

Chapter 3
Use and Occupancy

Learning Objectives

Upon completion of this chapter, you should be able to:

- Describe the classification system used by the model building codes to group buildings by use and occupant characteristics.
- Describe how multiple uses or mixed uses within a single building are addressed by the model building codes.
- Explain the need for special occupancy requirements in some buildings.
- Explain the fire official's responsibility in illegal changes in use.

INTRODUCTION

The consolidation of the three regional model code organizations into the International Code Council greatly simplified construction design and regulation in most of the United States. All 50 states, the District of Columbia and the Department of Defense have adopted the *International Building Code* as of July 2014, on a state or local level. In essence, structures that are designed or modified in the future will have to meet the requirements of the IBC. For most of the country, we now speak the same code language. An A-2 steak-house restaurant that seats 400 in Idaho is an A-2 in South Carolina, New Hampshire, Maryland, and New Mexico as well.

That wasn't the case prior to the development of the International Codes in 2000. The three regional model code groups, BOCA, ICBO and SBCCI, each produced a model building code. Although the codes had similar provisions in many areas, there were significant differences throughout the documents.

What the development of IBC did not accomplish, however, was to harmonize the construction classifications of *existing* buildings and structures. In the example of the 400-seat steak-house restaurant described above, what is classified as an A-2 under the IBC was classified as an A-3 under the *BOCA/National Building Code*, an A-2.1 under the *Uniform Building Code*, and an A-2 under the *Standard Building Code*. The existing buildings and structures within your jurisdiction were classified and constructed under specific editions of the codes adopted by your state or local governing body. These buildings must be maintained to that same level of protection, unless they are modified or significantly altered, or in the rare instance that a specific retrospective provision is enacted. Examples of retrospective requirements are laws that require retrofitting sprinklers in high-rises, hotels, hospitals and nursing homes and college dormitories.

Your road map through this part of the code journey is a copy of every edition of every code that has ever been adopted within your jurisdiction. The compulsive organizer who seeks to purge the office of unwanted clutter should be informed that codebooks are tools. Don't throw tools away, no matter how old they are. The time will come when you need them.

NOTE

The first job of the design professional is to classify every space within the structure based on the intended use and occupancy. It should also be your first consideration during an inspection of a building.

HOW TO USE THE BUILDING CODE

Design professionals are the primary users of construction codes. Code officials use the documents to assess compliance. The first two chapters of the IBC are largely administrative items and definitions. Technical provisions, or the nuts and bolts of the code, begin with Chapter 3, "Use and Occupancy Classifications." The first job of the design professional is to classify every space within the structure based on the intended use and occupancy. It should be your first consideration during an inspection of a building.

The **use group** classification of a building is probably the most significant single design factor that affects the safety of the occupants as well as the fire suppression forces that are called on in the event of fire. The building's height and size, type of construction, type and capacity of exit facilities, and fixed fire suppression systems are all dependent on this classification. The system of use group classification is the foundation for the building and fire prevention codes. As we will see, a mistake in laying the foundation can prove to be a disaster later on. Failure to maintain the foundation can also be disastrous.

One of the primary responsibilities of the building official is to verify, during the plan review process, that the design professional has accurately defined the use and occupancy of the building, and that the design of the building is in accordance with the code requirements for that use. The fire official's responsibility for the life of this same building is to determine that it is still operating under that specific use, and that the building protection features including means of egress, fire-resistance rated assemblies, and fire protection systems are being maintained. The fire official has the added responsibility of ensuring that functions and processes occurring within the facility are in accordance with the fire code.

USE GROUP CLASSIFICATIONS

The model building codes and NFPA 101/5000 contain ten general classifications for use and occupancy (Table 3-1). These are further subdivided by specific characteristics into groups, or use groups. Use and occupancy classifications are based on three issues:

1. The relative hazard of the processes or activities
2. The capacity of the occupants for self-preservation
3. Types and quantities of regulated materials

Group F (Factory/Industrial) is its own group based on the inherent hazards associated with industrial processes. Group I (Institutional) is based on the recognition that persons who are confined to hospitals, nursing homes and correctional institutions cannot readily escape from danger and need additional built-in safeguards. Group H (High-Hazard) recognizes the inherent danger associated with hazardous materials, regardless of the intended use. Group H occupancies are so classified based on the amounts of regulated materials on the premises, not what is being done with them.

Table 3-1. Use and Occupancy Classifications

IBC	NFPA 101, NFPA 5000
A—Assembly	Assembly
B—Business	Business
E—Educational	Educational
F—Factory or industrial	Industrial
H—High hazard	Health care
I—Institutional	Detention and correctional
M—Mercantile	Mercantile
R—Residential	Residential
S—Storage	Storage
U—Utility, miscellaneous, or special	Ambulatory health care
	Residential board and care

Note: Table 3-1 is not a table for comparison between the codes

Group A- Assembly

Group A is subdivided by the IBC into five different classifications based on function. The biggest hazard in Group A occupancies is the people, who are often unfamiliar with their surroundings. A common problem associated with unfamiliarity with a structure is that people feel safer exiting by the same route they entered. This can be deadly if main exits are under-sized or compromised. Additional hazards posed by large fuel loads, such as decorations, sets, scenery or acoustical treatments, entertainment, and the ingestion of alcohol by the patrons, can all increase the potential level of hazard.

Group B- Business

Group B occupancies are those in which business is conducted and services are provided. Group B is one of the least restrictive classifications, based on low fuel loads and relatively low occupant loads. It includes banks, barbershops, physicians' and dentists' offices, and post offices. It is also one of the widest ranging classifications, encompassing a broad spectrum of buildings from airport control towers to motor vehicle showrooms. Assembly occupancies established for use by fewer than fifty persons are considered Group B.

Group E- Education

Group E occupancies are for educational use in kindergarten through 12th grade, and day care for children older than 21/2 years of age. Colleges and adult education are classified as Group B or Group A depending on the number of occupants. Spaces used for trade and vocational education in high schools are classified based on the activities undertaken by the class. An area used for auto body repair training would be classified the same as an auto body shop—F-1.

Group F—Factory/Industrial

Group F occupancies include those used for assembling, disassembling, fabricating, finishing, manufacturing, packaging, repair or processing, provided they are not classified as Group H or Group S. Group F occupancies that store, handle, process, or use quantities of hazardous materials that exceed the permissible threshold, are classified as Group H. Aircraft repair hangars and motor vehicle repair garages are classified as Group S.

Group F is further subdivided into: Group F-1 Moderate Hazard and Group F-2 Low Hazard. Group F-1 occupancies are those that pose moderate fire hazards and involve combustible materials. Group F-2 is limited to noncombustible materials, which do not pose a significant fire hazard. Glass, metal, and ceramic manufacturing are examples. A key point to remember is that the packaging method used for the finished product must be considered when determining whether an occupancy is F-1 or F-2. Products packaged in more than a single thickness of combustible packing material are rightly classified as F-1, based on the fire hazard posed by the packing materials, even if they are noncombustible like pottery or glass.

Group H—High Hazard

Group H occupancies are those in which quantities of regulated materials exceed the permissible threshold established by the code. Previous editions of the International Codes as well as the legacy codes of the former model code groups, used terms such as *"exempt amount"* and *"permissible amount"* to describe the threshold quantities. Some found the terms confusing, assuming that use of the term "exempt" meant that the materials were exempt from the code. The IBC now refers to the threshold as the *"maximum allowable quantity."* If the quantity is exceeded, the occupancy must comply with the Group H requirements, provided it does not qualify for one or more of fifteen exceptions.

Group H-1 occupancies are those that contain over the allowable quantity of detonable material. Group H-2 occupancies are those with materials that pose a deflagration hazard or a hazard from accelerated burning. Group H-3 occupancies are those with materials that readily support combustion or pose a physical hazard. Group H-4 occupancies are those with materials that pose a health hazard. Group H-5 occupancies are facilities involved in semiconductor manufacturing and comparable research. The maximum allowable quantities are contained in tables in both the *International Building Code* and *International Fire Code*.

Group I—Institutional

Group I occupancies are those in which persons whose capacity for self-preservation has been diminished, whether by age, illness or infirmity, or incarceration (Figure 3-1). Age is only considered a limiting factor for children 2 ½ years of age or younger. Group I is subdivided into three use groups.

Figure 3-1. Group I-2 occupants are not capable of evacuating the building and require a higher level of protection referred to as a "Defend-in-place strategy." © Depositphotos.com

Occupants Group I-1 occupancies include assisted living facilities, halfway houses, and group homes for more than 16 occupants. A key characteristic of Group I-1 is that the residents require supervision but are capable of appropriate unassisted response to an emergency. Small assisted living facilities, halfway houses, and group homes with 16 or less occupants are classified as R-4.

Group I-2 includes hospitals, nursing homes, and detoxification facilities. The occupants are not **ambulatory** and are not capable of evacuating the facility. Group I-2 occupancies are designed to shelter or harbor the occupants within the structure by building extensive fire safety features.

Group I-3 occupancies are those that house persons under security or restraint, such as prisons, jails and detention centers. Group I-3 is further subdivided into five occupancy conditions with different levels of protection based on the level of confinement.

Group M—Mercantile

Group M occupancies are those in which stocks of goods, wares, and merchandise are displayed and sold, and are open to the public. The biggest hazard in Group M is the merchandise—the fire load. Mercantile occupancies sales of hazardous materials are regulated by limiting total quantities as well as container and package size.

Group R—Residential

Group R occupancies are those in which persons sleep and are not otherwise classified as Group I. The term residential typically connotes more activities than sleeping, such as eating, washing, cleaning, and preparing food. The key hazard or danger is that while sleeping, occupants are more at risk than during their waking hours.

Group R is statistically the most dangerous. An estimated 2,550 civilians died in residential fires in 2011. Of these deaths, 415 occurred in apartment fires; another 2,105 died in one and two- family homes. From 2009 to 2011, an estimated 3,045 civilian fire deaths were reported to fire departments across the country each year. An estimated 1,600 fatal fires in residential buildings occurred annually in recent years (2009 to 2011). These fires resulted in an annual average of approximately 2,495 deaths, 625 injuries and $188 million in property loss.[1]

Group R occupancies are subdivided into four use groups. Group R-1 occupancies are those in which the occupants are transient, such as hotels. Group R-2 includes apartments, dormitories, convents and vacation timeshares.

Group R-3 occupancies include detached one- and two-family dwellings and townhouses over three stories in height. The same structures, three stories or less in height, are considered beyond the scope of the IBC and are regulated by the *International Residential Code.*

Group R-4 occupancies are actually small I-1 facilities, housing up to 16 persons who need to live in a supervised environment, but are otherwise capable of appropriate unassisted response to an emergency.

Group S—Storage

Group S occupancies are used for storage and contain not more than the allowable quantities of regulated materials. Buildings that exceed the maximum allowable quantity threshold are classified Group H. The hazard is the fire load represented by the material and the storage arrangement. Group S is subdivided into S-1 Moderate Hazard and S-2 Low Hazard. The two hazard classifications are similar to those used in subdividing Group F.

Group U—Utility/Miscellaneous

Group U occupancies are those of an accessory character or miscellaneous structures such as barns, livestock shelters, private garages, tanks and fences over six feet.

MIXED USES

Buildings often contain multiple occupancies with different uses. This three-story office building (Figure 3-2) has a restaurant (assembly) and a computer store (mercantile) on the first floor and professional offices throughout the rest of the building.

Figure 3-2. Mixed-use office building. © Depositphotos.com

The model codes provide for these situations either by requiring that the whole building be constructed to all the requirements of the most restrictive use group, or by separating the areas with fire-rated assemblies, or by separating the building with fire walls. Separating with fire walls actually creates separate buildings. By far the cheapest, most attractive method of separating mixed uses is by using fire separation assemblies, but this method is sometimes impossible due to building height and area requirements, which we discuss in Chapter 4.

Incidental Use Areas

Incidental use areas are spaces that are incidental to the main use and have a higher level of hazard. Boiler rooms, parking garages, laundry rooms, waste and linen collection rooms, and laboratories are some examples. The IBC requires a fire separation, an automatic fire extinguishing system, or both, to isolate the incidental use area from the main occupancy.

Accessory Use Areas

An **accessory use area** is a portion of a building with a different classification than the main area, but which does not require a fire separation between the main area, due to its small size and low hazard level. Group H accessory use areas must always be separated. Accessory use areas of all other use groups are permitted to be non-separated, provided all accessory use areas combined are not more than 10 percent of the story in which it is located, and is within the height and area limits.

Once the use group has been determined, the code becomes a much friendlier document. Requirements that range from building height to the need for fixed fire protection systems are all based primarily on the use group of the building. The code writers found that basing these requirements on the intended use or the nature of the occupants of the building worked very well, with a few exceptions. Those few exceptions make up an entire chapter in your building code—Special Use and Occupancy.

SPECIAL USE AND OCCUPANCY REQUIREMENTS

For most buildings and structures, assigning a use group and then specifying building requirements for all buildings within that use group works relatively well. Most mercantile occupancies share common hazards. Most business occupancies have similar occupants and processes. But what if that business happens to be on the twenty-sixth floor of a high-rise building? Or what if the men's clothing store is in the middle of a giant shopping mall? The relative hazards suddenly change, and we begin comparing apples to oranges.

Building codes provide an enhanced level of protection for certain occupancies to compensate for special hazards, over and above those posed by the use of the building alone. The inherent hazards posed by being located twenty-six stories above the ground (Figure 3-2), or in a large open area with high fire loading such as a shopping mall, are addressed as special use requirements. The list of special uses seems to grow with each new edition of the model codes.

NOTE

Building codes provide an enhanced level of protection for certain occupancies to compensate for special hazards, over and above those posed by the use of the building alone.

Covered Mall Buildings

Understanding the concept of a covered mall is important. In your building code are definitions for *covered mall building, covered mall* and *anchor store or anchor building*. As with all of our subjects, read and understand the definition!

Figure 3-3. Pedestrian mall in a covered mall building.

A covered mall building is considered a building that encloses a number of tenants and occupancies such as retail stores, amusement facilities, restaurants, offices, and even transportation terminals where two or more tenants have a main entrance into the mall. The mall itself is the covered pedestrian walkway (Figure 3-3). It is located in the covered mall building.

Why all the fuss? Travel distances are measured within the mall itself, the pedestrian area, to the exterior or to an exit. Anchor buildings or anchor stores, which are usually large department stores, are not considered to be a part of the mall. The required means of egress from anchor stores are totally independent from the mall. Exits leading directly to the building exterior must be provided, as if that huge opening into the mall isn't there.

Occupancies that are part of the mall exit through the mall. But travel distance through the exit access is limited to 200 feet. Malls are large buildings with many areas over 200 feet from a door to the exterior, so exit passageways are provided. The rear corridors that mall managers refer to as "service and delivery" are actually the exits for the mall shops. The 200 feet of travel distance is to the entrance to the exit passageway, not to the outside. Consider those passageways as stair-towers turned on their sides. The merchants who stack empty boxes (see Figure 3-3) there, or block them with carts, or prop the doors open should be getting a visit from you.

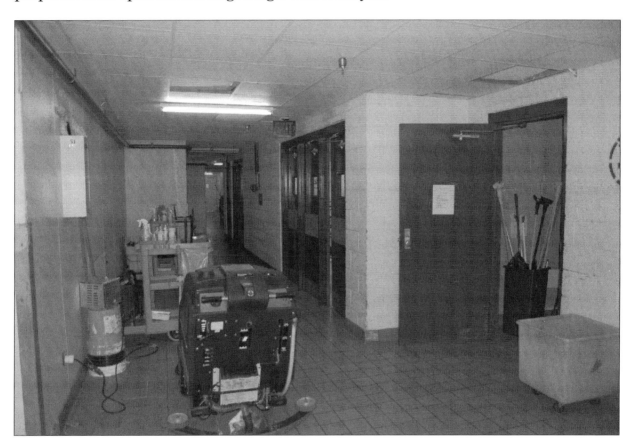

Fig. 3-3. Storing equipment in the exit or blocking fire doors open creates an unsafe condition

Nothing is permitted to be stored in an exit, period. Covered mall buildings have special requirements for egress, fire separation of tenants, fire suppression, and smoke control, over and above those of the use groups of the tenants.

High-Rise Buildings

High-rise buildings are those with occupied floors more than 75 feet above the lowest level of fire department access. The code attempts to address the practical difficulties faced by the occupants in fleeing the structure, as well as those faced by the fire department in manually extinguishing a fire. These include the use of noncombustible construction materials, structural ratings, installation of sprinklers, standpipes, and a fire command station with control of various building systems.

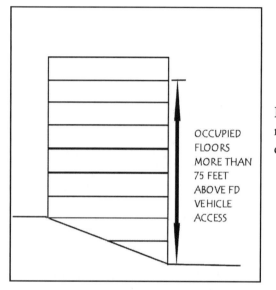

Figure 3-4. High rise buildings have occupied floors more than 75 feet above the lowest level of fire department vehicle access.

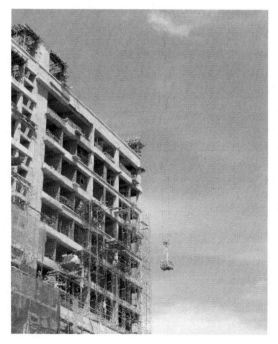

Fire-resistive construction (Figure 3-5), stairwell pressurization, selective alarming of floors, and stairwell locking requirements are all special requirements above and beyond those of the building's use group(s).

Figure 3-5. High-rise buildings are constructed of the most stringent fire-resistive construction type.

Special Amusement Buildings

Special amusement buildings range from fun houses at carnivals to laser tag facilities in strip shopping centers. The common element is that the means of egress is intentionally confounded through the use of a maze, movable partitions, darkness, sound effects, or a combination of all of them. Without special precautions, emergency egress would be severely impacted. The codes require the automatic restoration of adequate lighting and the stopping of special audio effects upon automatic smoke detection or activation of the sprinkler system. Constant attendance at a station with manual sound and lighting controls is also required. All decorative material must also be noncombustible.

Other Special Uses

Special requirements for building atriums, high-hazard occupancies, hazardous production material facilities (HPM), and institutional occupancies are also included in this chapter of your building code. For these limited situations, the use group classification system that works so well everywhere else in the code did not quite fit the bill. The code writers found they needed to add a little extra. In recent years, members of the International Code Council have begun to expand Chapter 4 to include provisions for motor-vehicle related, aircraft-related and ambulatory healthcare facilities. These changes help the code user locate requirements that in some cases were spread throughout previous editions.

CHANGES IN USE

Detecting and prohibiting illegal changes in use is a primary responsibility of fire inspector. Some changes in use do not appear that drastic on the surface, but some pose immediate hazards to the occupants. The 10,000-square-foot non-sprinklered retail store that turns into a bar by magic overnight might not seem like that big a change. At least not until you go out to the car and check your code book. The threshold for requiring an automatic sprinkler system in a retail store is 12,000 square feet, so the building was not sprinklered when it was originally built. Under the IBC, the threshold for requiring sprinklers in a bar or tavern is 5,000 square feet. The lack of sprinklers is not the only potential problem. The building will require panic hardware on the exit doors, a fire alarm system, additional exits, and more.

Illegal changes in use are a violation of the fire code. You have the responsibility and authority to prevent the illegal change. What you as the fire official do not have is the authority to order the required upgrades and approve the new use. Our retail store/bar example above clearly demonstrates the need for the fire official to work with the building official. Although the IFC authorizes both the fire official and building official to require the installation of sprinklers, that likely is not be the only deficiency requiring an upgrade to meet the requirements for the new use.

NOTE

Illegal changes in use are a violation of the fire code.

Changes in use require a complete analysis of a building prior to approval. It is as if the owner is building a new building on the site. The fire inspector who orders the installation of a sprinkler system and then finds out later that the building is too big, or built of the wrong type of materials, or built too close to the lot-line exposes himself and his jurisdiction to liability and public scorn, not to mention damaging the working relationship between the building and fire departments.

NOTE

No single other design element has a greater effect on the safety of the occupants than use group classification.

Summary

No single other design element has a greater effect on the safety of the occupants of a building than use group classification. The use of a building determines the number of occupants as well as their physical and mental characteristics. Their capacity for self-preservation as well as the physical hazards they will be exposed to are all addressed through the use group classification system. Special hazards and circumstances are addressed through special use and occupancy requirements.

The code builds on this classification, by regulating the height and area, type of construction, separation, and fire protection features of buildings based on use group.

Review Questions

1. In designing a building, the first and most critical step is to determine the _____.
2. Buildings in which multiple uses occur are called _____.
3. Describe three methods used by the model codes to address buildings in which multiple uses occur.

    a. _____
    b. _____
    c. _____

4. Name three types of buildings that have special use and occupancy requirements.

    a. _____
    b. _____
    c. _____

5. A(n) _____ is a large building on the exterior perimeter of a covered mall that has exits independent of the mall.
6. What impact do the packaging method and packing materials have on the Group F classification?
7. Under what circumstances can age be a determining factor in determining the use and occupancy of a building?
8. What is the principal occupant characteristic that results in different use group classifications for nursing homes and board and care facilities? _____
9. Are illegal changes in use a violation of the model fire codes? _____
10. Does a fire inspector have the authority to order the installation of a sprinkler system in a building that has been illegally changed from a warehouse to a nightclub? _____

Discussion Question

1. You are called by the administrator of an adult care facility classed as I-1 (IBC). She requests your assistance in updating the fire safety plan for her facility, because most of the occupants are in wheelchairs and the old plan called for total evacuation. She has drafted a new plan using the defend-in-place strategy in use at the last facility she worked at and would like your approval.

    a. Does the current occupancy reflect the use group assigned to the building?
    b. How does the use group affect fire safety and evacuation planning?
    c. Is the defend-in-place strategy proposed by the director compatible with the built-in fire safety features of the building?

KEY TERMS

use group

building code classification system whereby buildings and structures are grouped together by use and by the characteristics of their occupants

ambulatory

able to walk about without assistance; capable of sensing an emergency situation and appropriately responding by exiting the building

accessory use area

a portion of a building with a different use group classification than the main area, but which does not require a fire separation between the main area, due to its small size

Chapter 4
Building Limitations and Types of Construction

Learning Objectives
Upon completion of this chapter, you should be able to:

- Define fire resistance.
- Describe the five basic construction types.
- Explain why building height and area are regulated by the building codes.
- Describe the methods used by the codes to permit height and area increases.

INTRODUCTION

Two of the most effective methods to limit potential fire spread and prevent conflagrations have been limiting the size of buildings and regulating the materials used in construction. The same principle behind the city of London's 1189 ordinance requiring that all new buildings be constructed of stone walls with tile roofs[1] is used today in our modern building codes.

NOTE

The largest part of the model building codes addresses fire protection issues.

The primary purpose of a building code is to prescribe design and construction practices that will keep buildings from falling down. Besides gravity, there are many forces that act against a building. Snow loads, wind loads, and potential earthquake loads are all considered in the design and construction of a building by the building codes. The potential force that requires the most extensive code provisions is fire. The largest part of the model building codes addresses fire protection issues, fire safety, emergency egress, and structural stability when exposed to fire.

The key to understanding building code provisions for structural protection from fire is the concept of **fire resistance**. In simple terms it is the resistance of a building to collapse or to total involvement in fire. Frank Brannigan aptly referred to it as *fire endurance* in his classic 1971 *Building Construction for the Fire Service*. It is measured by the length of time typical structural members and assemblies resist specified temperatures.[2]

The term *fire resistance* is defined as that property of materials or their assemblies which prevents or retards the passage of excessive heat, hot gasses, or flames under conditions of use.[3] Fire-resistive construction elements, including methods used to test and assign relative values to their performance in hours, are discussed in Chapter 5.

TYPES OF CONSTRUCTION

We tend to make the subject of construction type much more difficult than it really is. There are three key points to remember when dealing with building construction types:

1. In everyday usage, the term "**combustible**" describes a material that will burn, and noncombustible describes a material that won't burn. It isn't quite that simple when it comes to construction regulation. For building materials to be classified as noncombustible, they must be tested in accordance with approved

fire tests. ASTM E 136, Test Method for Behavior of Materials in a Vertical Tube Furnace at 750°C, is used for elementary materials or those that are composed of a single product. Gypsum and concrete are elementary materials. For composite materials, such as gypsum board or sheetrock, the gypsum structural base is tested in accordance with ASTM E 136. The paper surface covering can be no thicker than 0.125 inches and is tested in accordance with ASTM E 84, Test Method for Surface Burning Characteristics of Building Materials. Composite materials that pass both tests are considered noncombustible materials, even though one component (the thin paper covering) will burn under fire conditions.

2. **"Protected"**, when applied to construction materials, means protected from the effects of fire by encasement. Concrete, gypsum, and spray-on coatings are all used to protect construction elements. When the code means "protected with a sprinkler system," it will say just that.
3. Having the ability to eyeball a building and determine the construction type is not a requirement of the job, and in fact will often be impossible for the inspector in the field—accept this. It is easier to determine what it is not than what it is, and usually that gets the job done.

An inspector needs the ability to recognize the questionable presence of combustible construction in a noncombustible building, or the apparent damage or compromise of rated components or assemblies.

> **NOTE**
>
> **"Protected", when applied to construction materials, means protected from the effects of fire by encasement.**

CONSTRUCTION MATERIALS AND FIRE SPREAD

The development and implementation of model construction codes in the first half of the twentieth century significantly reduced the conflagration potential throughout most of the United States. The conflagrations listed in Table 4-1 led to the development of building codes that severely restricted combustible building exteriors and roofs. The legacy model building codes provided for the establishment of **fire limits** or **fire districts** within cities and towns. Within fire districts or fire limits, combustible building exteriors and highly combustible roofs were prohibited. The concept exists today as Appendix D to the *International Building Code*, but is generally not adopted as part of the code.

Date	City	Fire Loss
June 1820	Savannah, Georgia	$3,000,000
December 1835	New York City	$17,000,000
December 1835	Charleston, South Carolina	$6,000,000
March 1852	New Orleans	$5,000,000
July 1866	Portland, Maine	$10,000,000
October 1871	Chicago (56 insurance companies bankrupted)	$168,000,000
November 1872	Boston (65 acres of the city)	$70,000,000
June 1889	Seattle	$5,000,000
November 1889	Lynn, Massachusetts	$5,000,000
October 1892	Milwaukee	$5,000,000
May 1901	Jacksonville, Florida	$11,000,000
February 1904	Baltimore (140 acres, 1343 buildings)	$40,000,000
April 1906	San Francisco (earthquake and fire)	$350,000,000
April 1908	Chelsea, Massachusetts (3500 buildings)	$12,000,000

Table 4.1 Source: Joseph K. Freitag, Fire Prevention and Protection as Applied to Building Construction, Wiley, New York, 1912, p.7.

Most of the brick buildings that line the streets of America's cities and towns can be traced to the establishment of fire districts during the twentieth century. Frame buildings were not permitted within fire districts, which generally encompassed the entire downtown area. Although conflagration potential has been significantly reduced, it has not been eliminated. Code provisions establishing fire districts were deleted from model construction codes or published as appendices with optional enforcement.

A March 26, 2004 fire that started on the site of a nearly completed student apartment complex near Virginia Commonwealth University in Richmond, Virginia, destroyed twenty-eight buildings (see Figure 4-1).

Figure 4-1 Fire spread from this four-story frame building and destroyed twenty-eight others. Frame construction was prohibited by fire limits provisions in previous editions of the building code. *(Photo courtesy of Dean Hoffmeyer/Richmond Times Dispatch)*

Two hundred firefighters and fifty pieces of equipment were involved in the incident. Damage exceeded 20 million dollars.[4] Fire was spread by flying brands created when combustible construction materials burned and were carried by 20 mile per hour winds. The student apartment building was a four-story frame structure that would have been prohibited if fire limits were still enforced within the city.

The use of fire limits or fire districts is unpopular among developers and design professionals because it has been estimated to add $8–$10 per square foot to the cost of a typical construction project. Many express the opinion that the widespread use of automatic sprinklers makes the concept obsolete.[5] Try telling people who were in downtown Richmond on March 26, 2004.

Five Construction Types

Five basic construction types are recognized by the model building codes. The terms vary a little between the different codes, but the concept is the same. Based on the classification from the 2012 *International Building Code* (Table 4-2).

Table 4-2 Fire resistance rating requirements for building elements (hours).

BUILDING ELEMENT	TYPE I A	TYPE I B	TYPE II A[d]	TYPE II B	TYPE III A[d]	TYPE III B	TYPE IV HT	TYPE V A[d]	TYPE V B
Primary structural frame[g]	3[a]	2[a]	1	0	1	0	HT	1	0
Bearing walls									
Exterior[f,g]	3	2	1	0	2	2	2	1	0
Interior	3[a]	2[a]	1	0	1	0	1/HT	1	0
Nonbearing walls and partitions Exterior	See Table 602								
Nonbearing walls and partitions Interior[e]	0	0	0	0	0	0	See Section 602.4.6	0	0
Floor construction and associated secondary members (see Section 202)	2	2	1	0	1	0	HT	1	0
Roof construction and associated secondary members (see Section 202)	1 ½[b]	1[b,c]	1[b,c]	0[c]	1[b,c]	0	HT	1[b,c]	0

For SI: 1 foot = 304.8 mm.

a. Roof supports: Fire-resistance ratings of primary structural frame and bearing walls are permitted to be reduced by one hour where supporting roof only.

b. Except in Group F-1, H, M, and S-1 occupancies, fire protection of structural members shall not be required, including protection of roof framing and decking where every part of the roof construction is 20 feet or more above any floor immediately below. Fire-retardant treated wood members shall be allowed to be used for such unprotected members.

c. In all occupancies, heavy timber shall be allowed where a 1-hour fire resistance rating is required.

d. An approved automatic sprinkler system in accordance with 903.3.1.1 shall be allowed to be substituted for 1-hour fire-resistance rated construction, provided such system is not required by other provisions of the code or used for an allowable area increase in accordance with Section 506.3 or an allowable height increase in accordance with Section 504.2. The 1-hour substitution for the fire-resistance of exterior walls shall not be permitted.

e. Not less than the fire-resistance rating required by other sections of this code.

f. Not less than the fire-resistance rating based on fire separation distance (see Table 602).

g. Not less than the fire-resistance rating as referenced in Section 704.10

Source: 2012 *International Building Code.* Copyright 2011, Washington, DC: International Code Council, Inc. Reproduced with permission. All rights reserved

Type I: Fire Resistive

In type I construction (see Figure 4-2), the structural elements are noncombustible and are protected. Type I is divided into two or three subtypes depending on the model code. The difference between them is the level of protection for the structural elements (expressed in hours). Structural steel is not exposed. A high-rise building with a steel structure encased in concrete is an example of a type I building.

Figure 4-2 Type I construction provides a high level of safety for those incarcerated in this jail.

Type II: Noncombustible

In type II construction (see Figure 4-3), the structural elements are noncombustible or limited combustible. Type II is subdivided into subtypes, depending on the level of protection (in hours) for the structural elements. The buildings are noncombustible, but are afforded limited or no fire resistance for the structural elements. A strip shopping center, with block walls, steel bar joists, unprotected steel columns, and a steel roof deck is a type II (unprotected) building.

Figure 4-3 Type II construction is noncombustible. Its structural components may or may not be protected with concrete, gypsum or another approved material, but it lacks the fire-resistance of Type I.

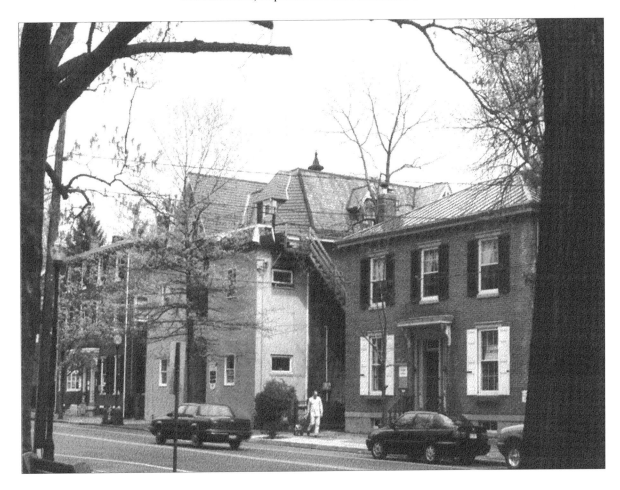

Figure 4-4 Type III "Ordinary" dates back to fire limits or fire districts contained in the legacy building codes.

Type III: Limited Combustible

(Ordinary) In type III construction (see Figure 4-4), the exterior of the building is noncombustible (masonry) and may be rated depending on the horizontal distance to exposures. The interior structural elements may be combustible or a combination of combustible/noncombustible. Type III is divided into two subtypes, protected and unprotected. The brick, wood joisted buildings that line our city streets are of type III or "ordinary construction." Buildings with a masonry veneer over combustible framing are not type III, they are sometimes Type II, but more often type V Frame buildings, where the veneer is not load bearing and hence, not structural.

Type IV: Heavy Timber In type IV construction, the exterior walls are noncombustible (typically masonry), and the interior structural elements are unprotected wood of large cross-sectional dimensions. Columns must be at least 8 inches if supporting a floor load, joists and beams a minimum of 6 inches in width and 10 inches in depth. Type IV is not subdivided. The inherent fire-resistive nature of large diameter wood members is taken into account. Concealed spaces are not permitted (see Figure 4-5).

Figure 4-5. This church is a Type IV building.

Type IV buildings feature noncombustible exterior walls and structural elements of large solid or laminated wood members. Log buildings are not Type IV.

Type V: Frame

In type V construction (see Figure 4-6), the entire structure may be constructed of wood or any other approved material. Brick veneer may be applied, but the structural elements may be wood. As the cost of lumber continues to rise, more and more composites and lightweight steel components are being used. Type V is subdivided into protected and unprotected, again depending on the protection provided for the various structural elements. Most residential structures are Type V.

Figure 4-6 Type V buildings may be constructed of any approved material.

FIRE RESISTANCE RATINGS

Each of the model codes and NFPA 220 has a table containing the ratings (in hours) for the various structural elements. The tables list required ratings by building component type, depending on the construction classification of the building. The construction classifications used by the model codes and NFPA 220 do not exactly match type for type. Table 4-3 provides an approximate comparison.

> **Note**
>
> **Each of the model codes and NFPA 220 has a table containing the ratings (in hours) for the various structural elements.**

Table 4-3 Comparison of the construction classifications of the model building codes.

IBC (2012) Table 601	IA	IB	IIA	IIB	IIIA	IIIB	IV	VA	VB
NFPA 220 (2015) Table 4.1.1	I 442	I 332	II 222	II 111	II 000	III 211	III 220	IV 2HH	V 111	V 000
BNBC (1999) Table 602	1A	1B	2A	2B	2C	3A	3B	4	5A	5B
SBC (1999) Table 600	I	II	IV 1 Hr	IV U	V 1 Hr	V U	III	VI 1 Hr	VI U
UBC (1997) Table 6-A	1	II FR	II 1 Hr	II N	III 1 Hr	III N	IV	V 1Hr	V N

Table 4-3 is a relative comparison gleaned from the 2012 IBC, 2015 NFPA 220, and the final editions of the legacy model codes.

Look at the table in your building code and compare the required rating for a column or structural frame member that supports more than one floor. Notice a big difference as you move from type I through type V Unprotected? How about for load-bearing exterior walls? We will see later that the code rewards fire-resistive design and penalizes the others.

HEIGHT AND AREA LIMITS

If the children's story *The Three Little Pigs* had been written by a fire inspector, it might have been a bit more of an action thriller. Instead of huffing and puffing and blowing the straw and stick houses down, the big bad wolf could have set a fire and let one house burn down the other. The third pig, living in his masonry house equipped with a sprinkler system would still live happily ever after, perhaps making a fortune in real estate by buying up the vacant lots left behind by the fire. He was a smart pig; after all, he built a type III building and his buddies built type V. He knew that all construction materials do not give the same performance.

There is an inherent risk associated with the type of materials used in the construction of buildings. Physical characteristics of the materials, including the ability to resist ignition and maintain structural integrity when exposed to fire, are what we have referred to as fire resistance. These characteristics are critical to the ability of the occupants to escape from the structure and, to limit fire spread to other structures, and for fire suppression forces to safely and effectively attack the fire.

> **Note**
>
> **Physical characteristics of the materials, including the ability to resist ignition and maintain structural integrity when exposed to fire, are what we have referred to as fire resistance.**

The model building codes use height and area limits to establish equivalent risks for structures based on four critical elements:

1. Use group
2. Construction type
3. Building frontage or open space that provides exposure protection and fire department access
4. The presence of an automatic sprinkler system

The *International Building Code* system of determining the allowable height and area of a building is similar to that used by the legacy building codes. IBC Table 503 lists the allowable number of stories and maximum area per floor based on the use group and construction type. The maximum area per floor listed in Table 503 is referred to as the tabular area, and is merely the starting point in the calculation.

> **Note**
>
> **The model building codes use height and area limits to establish equivalent risks for structures.**

Building Height

There are a couple of reasons that the model building codes require the vertical distance of a building to be measured. First is to determine whether the building meets the overall height limits set by the codes based on use group and type of construction. A second measurement is sometimes required by the code to determine whether the building must conform to the special requirements for high-rise buildings or requires the installation of standpipes.

The first edition of the International Building Code (2000) IBC simplified the matter of determining **building height** as being "the vertical distance from grade plane to the average height of the highest roof surface." Sounds easy, until you look at the definition of "grade plane" and just what is meant by "average height of the highest roof surface." To begin with, the **grade plane** is an imaginary plane representing the finished ground around the exterior walls of a building. If the ground level slopes away from the building, the grade plane is established using the lowest point within 6 feet of the exterior walls unless the lot-line is closer. The average height of the highest roof surface means top of a flat roof and the average height of the highest gable on a peaked roof.

In addition to height, the model codes regulate the number of stories permitted. The different codes use definitions of basement, first story, and story above grade to address portions of a building that are partially below grade and to determine whether a story is in fact a "basement" and not to be counted when determining whether the building meets the code limit on number of stories. This becomes especially important in modern multifamily residential buildings that are often of combustible construction and have large occupant loads (see Figure 4-7). The potential need for aerial and ground ladders and elevated streams gives new significance to how the area around a building is graded, and how well fire department access has been maintained.

Note

In addition to height, the model codes regulate the number of stories permitted.

In determining whether a building requires standpipes or must meet the high-rise requirements, a second measurement is taken from the lowest level of fire department vehicle access to the floor level of the highest occupied floor. This measurement is taken only on one side of the building—the lowest level that the fire department can drive equipment to. The measurement is made from the ground to the floor level of the highest floor that is occupied.

Figure 4-7 Grading and the installation of a decorative fence potentially affect the fire department from quickly accessing the building in the event occupants need to be rescued via ladders. © Depositphotos.com

Height Modifications

The building height and allowable number of stories specified in IBC Table 503 may be increased by one story and 20 feet in most use groups, if the building is equipped with an approved automatic sprinkler system. The codes require sprinkler protection for the entire building, not an area within the building, and the system must be designed to the appropriate sprinkler standard.

Area Modifications

The IBC defines building area as: "The area included within surrounding exterior walls, or exterior walls and fire walls, exclusive of courts..." The legacy codes have similar definitions. Area modifications could be described as changes to the footprint of the building or the actual amount of ground space it covers.

The tabular area from IBC Table 503 is increased if the building is equipped with an approved sprinkler system, or (and) if the building fronts open yards or public ways at least 20 feet in width. The installation of a sprinkler system results in a 300 percent increase for single story buildings and a 200 percent increase for multistoried buildings. The percentage of the building perimeter that faces open yards or public ways also results in an area increase. The advantage of sprinkler protection is obvious. Buildings that are situated at least 20 feet from interior lot lines and accessible by fire department vehicle have two fire protection advantages. The possibility of fire spread from or to surrounding buildings is significantly reduced by the open space. Fire suppression efforts are enhanced by the ability to access more of the exterior perimeter of the building.

Note

The ultimate area modification is included in each of the model building codes and is called the unlimited area building.

Unlimited Area Buildings

The ultimate area modification is included in each of the model building codes and is called the unlimited area building. It is generally limited to business, factory, mercantile, and storage occupancies and is fully sprinklered. Noncombustible construction or setbacks on all sides of up to 60 feet, or both, may be required.

EXPOSURE PROTECTION FOR BUILDINGS

The threat of fire spread to adjacent buildings and structures has been regulated by the model building codes in two ways. First, noncombustible and fire resistive construction are rewarded through the allowable height and area provisions of the code. Providing an open perimeter around the building enhances that reward. The second method is based on the distance from the building to the property line. The closer the building to its neighbors, the greater the exposure potential. The model building codes attempt to minimize fire spread to exposed buildings by increasing the required fire-resistance rating for exterior walls as the distance between buildings is reduced. Table 602 of the IBC lists ratings in hours based on distance, type of construction, and use group. Exterior wall protection and how we protect openings in exterior walls is discussed in Chapter 5.

Summary

Regulating the size of buildings based on the fire-resistive qualities of the construction materials used has been around for a long time. The National Board of Fire Underwriters' *National Building Code* contained provisions for height and area limitations and requirements for fire-resistive construction in 1905. These requirements provide for the safety of the occupants and establish exposure protection for adjoining structures. In establishing construction types, both combustibility and fire resistance of construction elements are considered.

Review Questions

1. _____ is the resistance of a building to collapse or total involvement in fire.
2. When used in describing construction types protected means _____
3. According to your building code, noncombustible unprotected construction is classified as type _____.
4. According to your building code, heavy timber construction is classified as type _____.
5. According to your building code, which types of construction must have noncombustible exterior walls? _____
6. What four elements are used by the model building codes to establish limits on building height and area?

    a. _____
    b. _____
    c. _____
    d. _____

7. Building height is determined by measuring from grade to _____ on a flat roof or _____ on a pitched roof.
8. List two reasons why the model building codes provide area increases based on the amount of a building's perimeter that faces a street or open lot.

    a. _____
    b. _____

Discussion Question

1. The management of a condominium apartment complex has applied for a permit to operate a place of assembly. A seating plan for the second floor of the community swim club building showing 287 folding chairs with adequate aisles and exits is attached to the application. Before leaving the office for the inspection, you review the file and find the following information:

    Community Swim Club Building

    1234 Any Street

> 2 Stories, Group B nonsprinklered
>
> Gross floor area 7200 sq. ft./ floor

   a. Is the building's use group classification appropriate for the intended use?
   b. What steps must the management take to legally establish a place of assembly?
   c. Is there anything about the building's size or construction that would preclude the establishment of a place of assembly in this building?

KEY TERMS

combustible

any material not classified as noncombustible

fire resistance

the resistance of a building to collapse or to total involvement in fire; property of materials and their assemblies that prevents or retards the passage of excessive heat, hot gases, or flames under conditions of use

noncombustible material

Materials which when tested in accordance with ASTM E 136:

1. Temperature of material rises ≤ 54°
2. No flaming after the first 30 seconds
3. If weight loss > 50%, no temperature rise and no flaming

protected

shielded from the effects of fire by encasement in concrete, gypsum, or sprayed-on fire-resistive coatings

fire limits or fire districts

designated areas within a jurisdiction where combustible construction is prohibited or restricted in order to reduce conflagration potential

building height

according to the IBC, "the vertical distance from grade plane to the average height of the highest roof surface"

grade plane

an imaginary plane representing the finished ground around the exterior walls of a building

David Diamantes

Chapter 5
Fire-Resistive Construction Elements

Learning Objectives

- Upon completion of this chapter, you should be able to:
- Describe the differences between specification codes and performance codes.
- Explain why a rated assembly must be maintained as a unit.
- Describe the origin of the standard time temperature curve, explain why it is still in use, and the significance of fire-resistance time values.
- List three nationally recognized testing laboratories involved in the testing of fire-resistance-rated assemblies.
- Explain why the labeling of certain rated assemblies is mandated by the codes.

INTRODUCTION

Fire walls, fire partitions, fire barriers, fire separation assemblies—all are designed to resist the spread of fire, but are they the same things? Fireblocking and draftstopping are installed to prevent the free passage of flame and products of combustion within concealed spaces. Is one as good as the other? Are they interchangeable?

When considering fire-resistive construction, a point to remember is that the model codes are for the most part prescriptive documents. They "prescribe" or state a required level of protection. There are a number of ways to meet the prescriptive requirements. Rather than specify the exact construction of a component such as a fire wall, the code identifies the performance features of the component. Where a **specification code** might require all fire walls to be of a certain thickness of masonry block, a **performance code** identifies the performance required (in hours). The design professional can use any **listed** assembly that provides that fire-resistance rating. The term "performance based design," is a different animal altogether. Prescriptive requirements are abandoned and buildings are designed to achieve desired levels of fire resistance and life safety through the use of computer models and mathematical calculations.

NOTE

The model codes are for the most part performance codes.

All fire-resistive assemblies required by the code will have performance requirements prescribed. The building designer will have specified the **assembly** (by design number) to be used to attain the required rating, or will have used one of the approved equivalent fire-resistance methods from the code. The IBC, UBC, and SBC contain specific provisions for structural protection. The important thing for the fire inspector to remember is that every rated assembly is composed of elements that are tested and listed as a whole. Whether a wall, a floor/ceiling assembly, or a door, the entire assembly is tested and listed together. If any portion of the assembly is left out or altered, the assembly is no longer the listed item, and more importantly, the assembly may not function as designed.

Examples of prescriptive or specification type code requirements, or requirements that are not performance based, are those that specify fireblocking and draftstopping materials:

Fireblocking shall consist of following materials:

1. Two-inch (51 mm) nominal lumber.

2. *Two thicknesses of 1-inch (25 mm) nominal lumber with broken lap joints. One thickness of 0.719-inch (18.3 mm) wood structural panel with joints backed by 0.719-inch (18.3 mm) wood structural panels.*
 3. *One thickness of 0.75-inch (19 mm) particle board with joints backed by 0.75-inch (19 mm) particle board.... (2012 IBC 718.2.1)*

Draftstopping materials shall not be less than ½-inch (12.7 mm) gypsum board, 3/8-inch (9.5 mm) wood structural panel, 3/8-inch (9.5 mm) particleboard, 1-inch nominal lumber, cement fiberboard, batts or blankets of mineral wool or glass fiber, or other approved materials adequately supported. (2012 IBC 718.3.1)

FIRE TESTS

Rated assemblies are subjected to tests specified in various standards including the National Fire Protection Association NFPA 251, Standard Methods of Tests of Fire Endurance of Building Construction and Materials, the American Society for Testing and Materials (ASTM) ASTM E119, and Underwriters' Laboratories (UL) UL 263. Check your building code to find the standards referenced.

Note

Every rated assembly is composed of elements that are tested and listed as a whole.

The components are assembled in the laboratory and then subjected to fire conditions in accordance with the **standard time temperature curve** (STTC) (see Figure 5-1). Acceptance criteria include:

- Ability to support a specified load

- Ability of the assembly to limit the transfer of heat to the unexposed side of the assembly

- Ability of the assembly to prevent the passage of heat or flame sufficient to ignite cotton waste

- Ability to prevent the passage of excess heat to steel members

- Ability to withstand exposure to a hose stream

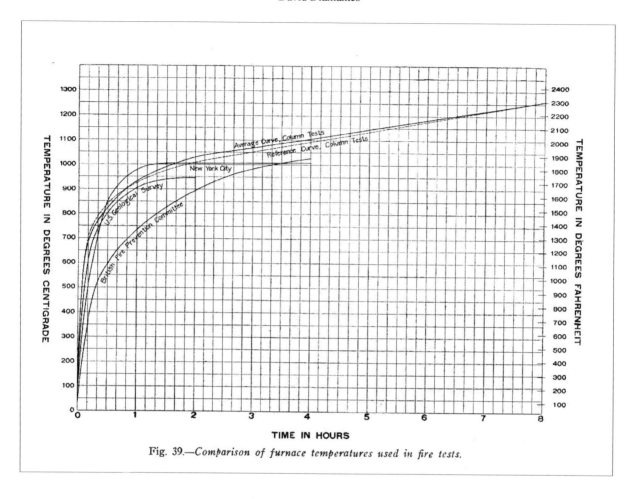

Fig. 39.—*Comparison of furnace temperatures used in fire tests.*

Figure 5-1. Standard Time-Temperature Curves in use in 1919. The U.S. Geological Survey curve is still used today. From *Fire Tests of Building Columns* (Chicago: Associated Factory Mutual Insurance Companies, et al, 1919) p. 114.

It is important to remember that the values (in time) assigned to rated assemblies are relative. A 2-hour wall survived the test protocol twice as long as the 1-hour wall did. When exposed to actual fire conditions, it is possible that neither one might hold back a fire for an hour. The STTC used to regulate the intensity of the "fire conditions" that our rated assemblies are exposed to are based on a fuel load of 8,000 BTUs per pound, which is about what you get from burning wood and paper products. The plastic materials, which are so evident in our modern buildings, weigh in at about 16,000 BTUs per pound.[1]

> **NOTE**
>
> **Components are assembled in the laboratory and then subjected to fire conditions in accordance with the standard time temperature curve.**

If you are wondering why we expose construction materials to test fires that do not necessarily mirror those found in modern buildings, you need to understand the origin of the standard time temperature curve. It has been around for a long time. Factory Mutual, the National Board of Fire Underwriters, American Society for Testing and Materials and the National Bureau of Standards was testing beams, girders, and columns at Underwriter's Laboratory in Chicago using the STTC in 1918.[2] At that time you could expect about 8,000 BTUs per pound from a fully loaded building of "ordinary" construction. It was a reasonable measure of performance that approximated fire conditions of the time. It is still a reasonable relative measure of performance. A relative measure of performance is really all a rating is.

> **NOTE**
>
> **Values (in time) assigned to rated assemblies are relative, and the "fire conditions" that our rated assemblies are exposed to are based on a fuel load of 8,000 BTUs per pound.**

RATED ASSEMBLIES

The fire-resistance ratings of building materials are not designed to give the occupant or firefighter a gauge as to how long a building will be tenable during fire conditions. They are performance comparisons under fixed fire conditions that do not mirror fire conditions in most modern buildings.

Accept the test results for what they indicate. Under laboratory conditions, the specimen performed to a given standard when exposed to a fire that did not replicate those found under actual fire conditions in today's environment. Do not expect to be able to eat your lunch on one side of a 1-hour fire separation wall because the fire next door has only been burning for 15 minutes.

> **NOTE**
>
> **Plastic materials, which are so evident in our modern buildings, weigh in at about 16,000 BTUs per pound.**

Perhaps the best-known source of listings is UL's *Fire Resistance Directory*. Other directories are produced by Warnock Hersey, the Gypsum Association, and others. Laboratories that perform tests of fire-resistive assemblies include the Factory Mutual

System, Inchcape Testing Laboratories, the Portland Cement Association, and Underwriters' Laboratories. Assemblies are specified on building plans by design number. By referring to the appropriate directory, a description and diagram of the assembly are available.

An important element that is often glossed over is the details crucial to the performance of the assembly. Fire barrier walls specify nail or screw schedules with fasteners of specific lengths. The edges must be staggered between layers of drywall, may or may not be required to be taped, and assemblies are sometimes product specific (see Figures 5-2 and 5-3). When it comes to rated assemblies, the details are everything.

NOTE

When it comes to rated assemblies, the details are everything.

WALLS AND INTERIOR PARTITIONS, NONCOMBUSTIBLE			
GA FILE NO. WP 1716	GENERIC	2 HOUR FIRE	40 to 44 STC SOUND
GYPSUM WALLBOARD, STEEL STUDS			

Base layer 5/8" type X gypsum wallboard or gypsum veneer base applied parallel to each side of 3 1/2" 20 gage steel studs 24" o.c. with 1" Type S-12 drywall screws 12" o.c. **Face** layer 5/8" type X gypsum wallboard or gypsum veneer base applied parallel to each side with 1 5/8" Type S-12 drywall screws 12" o.c. Studs attached to each side of floor and ceiling runners by welding or with 1/2" Type S-12 panhead screws.

Joints staggered 24" each layer and side.

Bracing: Lateral bracing spaced not over 40" o.c. shall be 1" by 18 gage steel straps attached to each side or channel bracing attached to each stud with a clip angle. For studs with holes or punch-outs in the web the "Q" factor shall be determined by means of stub column tests. Tested at 80 percent of design load. **(LIMITED LOAD-BEARING)**

Thickness:	6"
Limiting Height:	Subject to design
Approx. Weight:	10 psf
Fire Test:	UL NC 506-6, 7-29-82, UL Design U425
Sound Test:	See WP 1615 (NGC 2250, 1-3-68)

Figure 5-2 Specifications for a 2-hour fire barrier wall. (*Courtesy of the Gypsum Association, Washington, D.C.*)

In Chapter 4, we discussed the model code tables that list fire-resistance ratings for the five construction types. Conformance with these requirements can sometimes only be determined through the use of the listing documents and approved plans for the building.

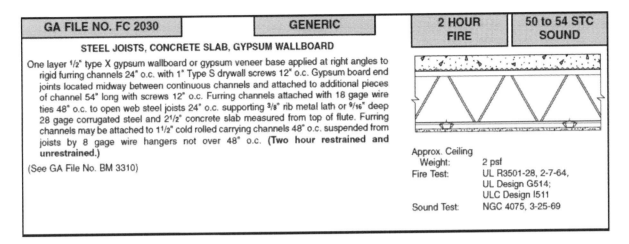

Figure 5-3 Specifications for a 2-hour floor-ceiling assembly, (*Courtesy of the Gypsum Association, Washington, D.C.*)

Maintenance of Rated Assemblies

The life of a rated assembly is often short and full of danger. Plumbers, telephone and cable installers, and electricians seem driven to run their services through fire separation assemblies. New building owners aren't happy with the listed locks on the rated stairwell doors and frequently install non-listed hardware. The lawyer has a residential mail slot installed in his rated office door. The employees at the auto body shop go one step further from propping open the rated exterior door—they also remove the self-closer. There is no level of fire protection that can be put into a building that a big enough hammer cannot take out.

Ensuring the maintenance of fire-resistance rated assemblies is the responsibility of the fire inspector. Damage can be easily overlooked and often will be missed, sometimes by several inspectors over the years. The cry from the business owner that "This is the first time in five years that anybody mentioned it," doesn't make the potential hazard to the occupants and business owner any less. The rating must be restored.

The 2012 *International Fire Code* section 703.1 is very specific. "*The required fire-resistance rating of fire-resistance-rated construction (including walls, fire-stops, shaft enclosures, partitions, smoke barriers, floors, fire-resistive coatings and sprayed fire-resistant materials applied to structural members and fire-resistant joint systems) shall be maintained. Such elements shall be properly repaired, restored, or replaced when damaged, altered, breached, or penetrated.*" The other model fire codes have similar requirements. The 2015 *Uniform Fire Code* section 12.3.3.1 adds "*removed or improperly installed*" to the mix.

Labeling is an important safety element of the code. Fire doors, fire windows, fire shutters, and fire dampers all must bear the **label** of an approved testing agency. It shows the inspector that construction elements are listed for a particular application, and shows the contractor that his supplier is delivering the proper materials. Labeling costs money. Some manufacturers offer the same door with or without the label. The only difference is that one has a small embossed piece of metal attached.

The thought of saving a few dollars each on the hundreds of doors needed for a large apartment project by buying unlabeled units might appeal to the builder who is trying to maximize his profit. Same door or not, they are unacceptable without the label. The code is specific. The required labels are marks that will be checked by inspectors for the life of the building. Treat the label as a vital element of the assembly. It must be there.

Fire Walls versus Fire Barriers

In order to understand the difference between the various structural components that are designed to prevent the spread of fire from one building to another; or one area within a building to another, go back to the building code definitions. The different codes have had their own terms for years. In formulating the first edition of the *International Building Code*, the members BOCA, ICBO, and SBCCI had to agree on common terms for the new building code. One thing should be clear to every person who has reason to use any of the codes or referenced standards—*read the definitions*.

> **NOTE**
>
> **Fire doors, fire windows, fire shutters, and fire dampers all must bear the label of an approved testing agency. Treat the approved testing agency label as a vital element of the assembly; it must be there.**

Fire walls are designed with the structural stability to be able to withstand collapse of the structure on either side and remain standing (see Figure 5-4). The wall does not have to be designed on the assumption that fires could start on both sides at the same time. The wall can rely on the structural stability of the buildings on either side, but must be designed with break-away bracing so that either side can collapse without causing the fire wall to collapse.[3] Fire walls may have parapets, depending on the type of construction and combustibility of the roof, and may be required to extend beyond the building's exterior walls.

Figure 5-4

This fire wall will separate single-family townhouses. It is being constructed in accordance with the Gypsum Association's design manual and is rated for 2-hours.

NOTE

A fire wall is designed with the structural stability to be able to withstand the collapse of the structure on either side.

Fire barriers, fire partitions, and fire separation walls are assemblies that separate and subdivide spaces within a building. The IBC mandates the use of a fire barrier to separate exits and exit passageways from the rest of the building, to separate mixed uses, individual fire areas, and to separate incidental use areas like boiler rooms. Fire barriers are rated assemblies that extend from the top of the floor/ceiling below to the underside of the floor/ceiling or roof deck above. They extend through any concealed spaces, such as ceilings, even if they are rated.

Fire partitions are used to separate dwelling units, guest rooms in hotels, tenant spaces within covered malls, and for corridor wall construction. With the exception of corridor walls, fire partitions are rated but are permitted to terminate at a rated ceiling.

NOTE

Fire separation walls, fire barriers, and fire partitions are assemblies that separate and subdivide tenant spaces within a building; fire walls separate buildings.

Fire walls separate buildings. Fire barriers and partitions separate or subdivide areas within a single building. Generally walls that separate occupancies within strip shopping centers are not fire walls. There is a very good chance that the strip shopping center where you buy your groceries, pick up your dry cleaning, and rent videos is one big building with multiple tenants. The tenant separation walls may be fire barriers depending on the use groups involved and the type of construction used. More often than not, they are nonrated partition walls used solely for security purposes. This is an important issue during a fire incident. Fire officers who assume that tenant separation walls extend to the roof deck, or have a fire-resistance rating, may be in for a surprise. The total area within the four exterior walls may not have a single rated element within the entire shopping center.

Another significant issue involves the handling of regulated materials using the fire codes. Maximum allowable quantities for materials are per building, not per tenant space. We'll discuss this in Chapter 12 *Hazardous Materials*.

Exterior Opening Protectives

As you recall from Chapter 4, type of construction is one of the factors that determines the required fire-resistance rating for the exterior walls of a building. The other factors are use group and fire separation distance. The 2012 IBC defines fire separation distance as:

The distance measured from the building face to one of the following:

1. *The closest interior lot line;*
2. *To the centerline of the street, an alley or public way; or,*
3. *To an imaginary line between two buildings on the same property. the distance shall be measured from the face of the building.*

Figure 5-5. This fire shutter provides exposure protection. Courtesy of Lionel Duckwitz.

Depending on these factors, the exterior of the building, including openings such as doors and windows, may have to be provided with specific fire-resistance ratings (see Figure 5-5). The number and size of openings may be limited, or in some cases prohibited. Ratings may be reduced if the building is fully sprinklered. The intent of the codes is to limit the spread of fire from one building to those adjacent. The rating required for the **opening protectives** is not typically the same as for the walls themselves. An exterior wall that requires a 2-hour fire-resistance rating under the *International Building Code* would require 1 ½-hour-rated opening protectives (716.5). In all cases, exterior openings that require protection must be provided with a minimum of ¾-hour rated assemblies.

NOTE

The rating required for the opening protectives is not typically the same as for the walls themselves.

Fire Door Assemblies

The most abused fire-resistance-rated assembly is the fire door, simply because it has moving parts and comes in contact with humans more than any other assembly. Fire doors come in a multitude of shapes and sizes. These include single side swinging, side swinging doors in pairs, Dutch doors, and horizontal and vertical rolling fire doors.

NOTE

The most abused fire-resistance-rated assembly is the fire door, simply because it has moving parts and comes in contact with humans more than any other assembly.

The assembly (see Figure 5-6) includes the door, frame, and all of the hardware including the hinges, latches and locks, closing mechanism, and any special features such as fusible links. These assemblies are tested and listed as a whole. They must be installed and maintained as a whole. The installation of locks, hold-open devices, or other special features after the fact, void the listing and violate the integrity of the rated assembly, unless the additional items are themselves listed for installation on that door and are installed in accordance with the manufacturer's specifications. As with exterior opening protectives, the required rating for fire doors in a rated wall or partition may be less than the wall itself. Each of the building codes has a table that lists the minimum opening protection for the various rated wall assemblies.

Figure 5-6. The fire door assembly includes the door, frame, hardware and locks and latches. Figure 5-1 shows an illegal, kick-down, hold-open device and holes from a previous hardware installation that should be repaired.

Structural Members

The ratings of columns, girders, trusses, beams, and other structural members are provided by a listed method such as encasement in concrete, a gypsum assembly, or with a spray-on encasement product (see Figures 5-7, 5-8 & 5-9). The maintenance of this protection can be difficult if the areas are subject to pedestrian or vehicular traffic, industrial activity, or tradesmen who decide that the assembly is "in the way" of their cable or pipe installation.

Figure 5-7 Spray-applied coatings provide fire resistance for steel building components. If this column is not enclosed within a wall assembly, impact protection must be provided.© Depositphotos.com/Alex Grichenko

Impact protection is required by the building codes and is provided for columns by the installation of corner guards or a substantial jacket of metal or other noncombustible material. Column impact protection should be treated by the fire inspector as an integral part of the assembly. He should order any damage to the guards or metal jacket to be repaired. If the impact protection method is ineffective, the owner should be ordered to provide an adequate method. Care should be taken to ensure that any new method of protection does not damage the rating of the structural member.

NOTE

Column impact protection should be treated by the fire inspector as an integral part of the assembly.

Figure 5-8. Sheetrock has been added to these columns for aesthetics and to protect the spray-on fireproofing. from impact damage.

Figure 5-9. Spray-on coating provides fire-resistance for steel building components. Courtesy of Duane Perry

Fireblocking and Draftstopping

Fireblocking and draftstopping are often taken for granted until a room and contents fire burns off the roof of an apartment building or the crew that was opening up windows on the floor above the fire reports that they are now operating a hose line. Draftstopping is designed to prevent fire spread within large concealed spaces such as attics and floor assemblies that use web trusses or have suspended ceilings.

> **NOTE**
>
> **Fireblocking and draftstopping are often taken for granted until a room and contents fire burns off the roof of an apartment building.**

Fireblocking is used to prevent the spread of fire within small concealed passages such as floors, walls, and stairs. The maintenance of these features is often complicated because access to concealed spaces is limited. Often the only persons who enter are tradesmen who run a new half-inch electrical cable through a 6-inch square hole, or building occupants who decide that void spaces make great storage spaces.

The requirement for the draftstopping of attic spaces has changed over the years. The 1970 BOCA Basic Building Code required firestopping in the attic every 3,000 square feet unless the roof and attic were noncombustible. By 1981, draftstopping was required in line with tenant separations unless the building and attic were sprinklered.

Fireblocking is installed at ceiling and floor levels of combustible construction, as well as at connections between horizontal and vertical concealed spaces. Openings around vents, ducts, chimneys, and fireplaces are also fireblocked.

Fire Dampers

Fire dampers are usually required to be installed wherever air distribution systems penetrate a rated assembly. They must be listed for the particular application and labeled. Like fire doors, the fire damper rating and the rating of the penetrated assembly are not always the same. The model codes permit 1 ½-hour fire dampers in 2-hour assemblies. (See Figure 5-10). Listed opening protectives must be installed in accordance with manufacturer's instructions.

Figure 5-10. This fire damper failed to operate because a contractor used a fastener that was too long. (See the white circle.) One leaf closed. But the other caught on the fastener and remained open, allowing smoke to enter an adjacent apartment.

Access for maintenance and inspection of fire dampers is required by the building codes. If you cannot get to it to inspect it, then a technician cannot get to it to maintain it. Lack of access is a serious problem that must be dealt with to ensure the integrity of the rated assembly.

Other Rated Assemblies

Other fire-resistance rated assemblies, including fire windows, fire shutters, wired glass panels, and other opening protectives (see Figure 5-11) all must be maintained in accordance with their listings by the owner of the building or structure. Fire inspections should include a check of testing and maintenance records, as well as a visual check of the assemblies themselves. A thorough inspection requires a look above that false ceiling every so often, and a good look at any new utility work. Snoop around a little; good inspectors are nosy about the building they are inspecting.

Figure 5-11. This opening protective protects a 1-hour roof ceiling assembly. The contractor was surprised when his neatly framed opening with a drop-in piece of ½" plywood was rejected during the final inspection, and scrambled to install the proper protection. Courtesy of Howard Bailey.

Summary

The building code requirements for fire-resistant construction features are generally prescriptive code rather than performance code requirements. They prescribe the required performance in hours. Requirements for structural elements vary according to the type of construction and can be found in the model codes in table form.

Assemblies are tested under fire conditions according to the test protocols using the standard time temperature curve. NFPA 251 for building materials and NFPA 252 for fire doors are examples of the test standards. Assemblies are tested and listed in directories by Underwriters' Laboratories, Warnock Hersey, the Portland Cement Association, and other approved testing agencies. The rating (in hours) should not be considered an absolute indication of how long the assembly will withstand actual fire conditions, because the test protocol does not replicate actual fire conditions in modern buildings.

Assemblies must be listed for the application, and some assemblies such as fire doors, fire dampers, fire shutters, and fire windows must bear the label of an approved testing agency.

Review Questions

1. For the most part, the model building codes are considered to be _____ type codes.

2. A code provision that requires all fire walls to be constructed of a certain thickness of masonry would be considered a _____ code provision.

3. Rated assemblies and components are tested under fire conditions prescribed by the _____.

4. Name three organizations that list fire-resistance-rated assemblies.

    a. _____

    b. _____

    c. _____

5. List two fire-resistance-rated assemblies that must bear a label.

    a. _____

    b. _____

6. What is the primary difference between a fire wall and a fire barrier?

7. _____ is designed to prevent fire spread within large concealed spaces such as attics.

8. Wherever air distribution systems penetrate a rated assembly, a _____ must be installed.

9. Fire-resistance ratings for listed assemblies are expressed in terms of _____.

10. Is the fire-resistance rating of an assembly a good indicator of its actual performance under fire conditions? _____.

Discussion Question

1. During your inspection of a plumbing supply warehouse, the owner is proud to inform you that business has been so good that he has expanded. He has purchased the building next door and doubled the size of his operation, expanding his 10,000-square-foot warehouse into 20,000 square feet by making two 10-foot openings in the block wall that separated the two nonsprinklered occupancies.

   a. What are some of the code provisions that address this expansion?

   b. What should the owner have done prior to expanding?

   c. What actions would you take in this case?

KEY TERMS

specification code

a code that specifies a type of construction or materials to be used

performance code

a code that assigns an objective to be met and establishes criteria for determining compliance, e.g., requirements for fire assemblies rated in "hours" and building materials rated as "noncombustible" when tested to specific protocols

listed

equipment or materials included in a document prepared by an approved testing agency indicating that the equipment or materials were tested in accordance with an approved test protocol and found suitable for a specific use

assembly (rated assembly)

a building component such as a door, wall, damper, or ceiling composed of specific parts and tested and listed as a unit

standard time temperature curve (STTC)

curve representing the standard reproducible test fire used since 1918 to measure the fire resistance of building materials

label

a permanent identification affixed to a product by a manufacturer indicating the function and performance characteristics of the product, name of the manufacturer, and the name of the approved testing agency that tested a representative sample of the product

opening protective

a rated assembly such as a door or window that provides a protected opening in a rated wall or partition; there are interior as well as exterior opening protectives

Chapter 6

Installation of Fire Protection Systems

Learning Objectives

Upon completion of this chapter, you should be able to:

- List the four classes of fire and give examples of each.
- Identify fire suppression system agents that are compatible with the different classes of fire and describe the method of extinguishment provided by each.
- List and describe the four basic types of sprinkler systems.
- Describe the differences between the NFPA 13, 13D, and 13R sprinkler standards and the application of each.
- Describe the function of standpipe pressure-reducing valves and the potential hazards posed by improper installation.
- Describe the purpose for the development of the UL-300 standard for restaurant cooking equipment.

Introduction

Fire protection systems is a broad term that includes built-in extinguishing systems such as automatic sprinklers, gaseous agent fire suppression systems, detection and alarm devices, smoke control and removal equipment, fixed protection for cooking operations and industrial equipment, and portable devices such as fire extinguishers. Each is designed for a specific hazard and is installed in accordance with the code to address that hazard. If the hazard changes, the fire protection system must be reevaluated to determine its potential effectiveness given the new condition.

Fire-Extinguishing Systems

Systems that automatically control or extinguish fires have become commonplace in our world. The cheeseburger you choked down at lunch today was hopefully cooked on a grill that was installed as required by code, under an approved exhaust hood and duct system. If there had been a fire on the grill while your order was cooking, it would probably have activated a wet chemical extinguishing system, suppressing the fire and shutting off all of the protected cooking appliances. You might not have eaten that cheeseburger.

> **NOTE**
>
> **If the hazard changes, the fire protection system must be reevaluated to determine its potential effectiveness given the new condition.**

Although you might not have noticed, the restaurant was also probably protected by an automatic sprinkler system. Most people do not pay attention to sprinkler heads—they have become part of the scenery people expect in a modern building.

Classification of Fires and Extinguishing Agents

Fires are classified according to the material or fuel involved. Fire-extinguishing agents are rated according to this classification system to ensure that the agent is compatible with the hazard. This is important for two reasons. First, the physical properties of the agent must match the hazards presented. Discharging a 2 ½-gallon water extinguisher onto a burning circuit breaker panel would definitely be an example of the use of an incompatible agent. Because water conducts electricity, the user would be at considerable risk of electrocution, and the fire would not be extinguished.

> **NOTE**
>
> **The physical properties of a fire-extinguishing agent must match the hazards presented.**

Our second consideration regarding agent compatibility is between agents themselves. Some extinguishing agents, notably different dry chemical compounds, can react with one another, diminishing their effectiveness. Table 6-1 includes examples of the five classes of fire, with appropriate extinguishing agents.

Table 6-1 *Classification of fires and extinguishing agents.*

Fire Class	Description	Examples	Extinguishing Agents
A	Common combustibles	Wood, paper, cloth, grain	Water, dry chemical, Foam, clean agent and some halon
B	Flammable liquids and gases	Gasoline, oil, grease, LPG	CO_2, dry chemical, clean agent and halon, foam
C	Energized electrical equipment	Energized class A material such as a household appliance	CO_2, dry chemical, Halon
D	Combustible metals	Magnesium, sodium, potassium	Dry powder
K	Cooking appliances with combustible media	Deep fat fryers	Wet chemical

Extinguishing Fires

In order to understand the various methods of fire extinguishment, an explanation of fire is in order. This seems elementary, but fire investigators spend extended periods on the witness stand establishing their credibility as expert witnesses. They are routinely asked for a definition of fire. Some of the definitions published in texts over the years go to the point of stating *fire is only sometimes accompanied by the evolution of light and heat in varying degrees.*

Since we are not on the stand, we will settle for the definition that I remember from my days in recruit firefighter training. **Fire** *is the rapid oxidation of a material, accompanied by the evolution of light and heat.* Fire is a chemical chain reaction. To put out the fire, you must stop the reaction. The older texts on fire prevention and firefighting explained the reaction by using a triangle (see Figure 6–1). The three sides are heat, fuel, and oxygen.

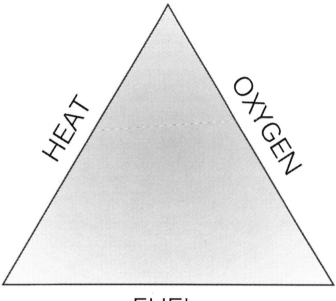

Figure 6-1. The three sides of the Fire Triangle

By removing any one of the sides, the reaction is halted and the fire is extinguished. Remove the heat by cooling the fire with a hose stream; deprive it of oxygen by blanketing the fire with carbon dioxide; or deplete its source of fuel say, by shutting down the leaking gas line, and the fire will be extinguished. All three methods have proven effective and are the basis for most modern fire extinguishing systems.

The old fire triangle is also the basis for many of the provisions of our modern fire codes. Regulations for the storage and handling of materials and for the use of open flames and heating appliances are all geared toward keeping the three sides of the triangle effectively segregated from one another.

NOTE

The old fire triangle is also the basis for many of the provisions of our modern fire prevention codes.

Modern research led to the update of the fire triangle. An extra element, the molecular chain reaction, was added to create the fire tetrahedron (see Figure 6-2). With the new element came another method of fire extinguishment—interruption of the molecular chain reaction. Certain extinguishing agents such as dry chemical, clean agent and halon extinguish fires by this method.

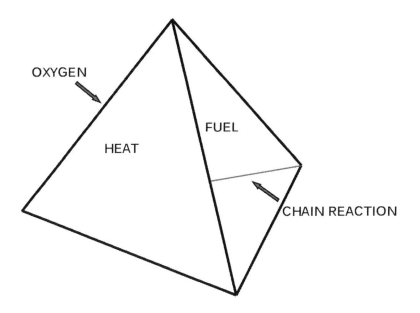

Figure 6-2 The Fire Tetrahedron ads a fourth leg, the chemical chain reaction to the Fire Triangle

Automatic Sprinkler Systems

By far the most prevalent, economical, and efficient fire protection system in use today is the automatic sprinkler system. The advantages over other systems are significant. What other system uses an extinguishing agent that is cheap, plentiful, nontoxic, and usually supplied to the premises by the municipality?

> **NOTE**
>
> **By far the most prevalent, economical, and efficient fire protection system in use today is the sprinkler system.**

Consider this:

- Ordinary combustibles produce about 8,000 BTUs per pound when fully involved in fire. One pound of water is capable of absorbing nearly 1,000 BTUs (970.3 to be exact).[1] One gallon of water (8.33 pounds) can absorb over 8,000 BTUs, so one gallon of water can absorb the amount of heat produced by one pound of burning Class A combustible material.
- Sprinkler systems activate quickly, activating an alarm and discharging water only on the fire area. Water damage is considerably less than when fires are manually extinguished by firefighters, minimizing business interruption.
- Sprinkler systems are in service around the clock and are fully automatic.

History and Development

The forerunners of the modern automatic sprinkler system were perforated pipe and open sprinkler systems installed in mill properties from about 1850 until 1880. Neither system was automatic and relied on supply valves being manually operated in the event of fire. Application at the seat of the fire was generally insufficient and water damage was extensive. Since all the discharge orifices in both systems were open, the entire area covered by the system was equally wetted.

In 1878, the first truly automatic sprinklers were installed. Technological improvements quickly followed, led by Frederick Grinnell of the General Fire Extinguisher Company.[2] The company, now bearing Grinnell's name, is still prominent in the sprinkler industry today.

NFPA 13, *Standard for the Installation of Sprinkler Systems* was first printed in 1896. Since that time it has been regularly updated and expanded. Technological advances have necessitated the breakdown of the original sprinkler standard into multiple standards.

Sprinkler Standards

The demand for lower cost sprinkler systems to protect one- and two-family dwellings and smaller multifamily buildings led to the development of NFPA 13D and NFPA 13R (see Table 6-2 below). Inspection, maintenance, and testing provisions for all water-based fire protection systems were compiled into a single standard—NFPA 25. The IBC, BNBC, and SBC legacy codes adopted the NFPA sprinkler standards by reference, the UBC legacy code adopted NFPA 13 and 13R by transcription as UBC Standards 9-1 and 9-3, respectively.

Table 6-2 *Sprinkler Standards Referenced by the Model Codes*

NFPA 13	*Standard for the Installation of Sprinkler Systems*
NFPA 13D	*Installation of Sprinkler Systems in One- and Two-Family Dwellings and Manufactured Homes*
NFPA 13R	*Standard for the Installation of Sprinkler Systems n Low-Rise Residential Occupancies*
NFPA 25	*Inspection, Testing and Maintenance of Water-Based Fire Protection Systems*

The 1999 edition of NFPA 13, *Standard for the Installation of Sprinkler Systems* was expanded to include sprinkler requirements for hazards such as rack and pile storage and special hazards such as rolled paper, plastic commodities, baled cotton, and rubber tires that were previously contained in other NFPA standards.

Additional sprinkler requirements are also found within NFPA standards that pertain to specific processes and occupancies. NFPA 30, *Flammable and Combustible Liquids Code* and NFPA 30B, *Manufacture and Storage of Aerosol Products* are examples. Each of the standards provides a different level of protection to address a specific hazard or hazards. The notion that all sprinklered buildings are equally protected, regardless of the contents or system design, is a dangerous one, embraced unfortunately by too many in the fire and building regulation communities.

NOTE

The notion that all sprinklered buildings are equally protected, regardless of the contents or system design, is a dangerous one.

NFPA 13 Systems

NFPA 13 systems are by far those most commonly found installed in buildings in the United States. NFPA 13, *Standard for the Installation of Sprinklers* is the direct descendent of the original sprinkler standard developed by NFPA in 1896. As we learned in Chapter 1, the existence of nine radically different standards for sprinkler pipe size within 100 miles of Boston was a primary reason for the establishment of the National Fire Protection Association. NFPA 13 has been actively maintained by the organization and used throughout the world for over 100 years.

NOTE

The NFPA 13 sprinkler system is a property protection system that has had remarkable success at protecting people by quickly suppressing and often extinguishing fires in their incipient stages.

The NFPA 13 sprinkler system is a property protection system that has had remarkable success at protecting people by quickly suppressing and often extinguishing fires in their incipient stages. Heat from the fire rises to the ceiling where sprinklers are installed and act as combination heat detection and fire suppression devices. Once the operating elements of the sprinkler head (generally a solder link or frangible bulb) reach their operating temperature, they fail and water is discharged onto the fire through the open orifice.

Standard automatic sprinklers are designed to operate at specific temperatures. Ordinary temperature-rated heads, which operate at between 135°F and 170°F, are required to be installed throughout buildings, unless the area is subject to high heat conditions. Boiler rooms, skylights, attic spaces, and areas above certain machinery and equipment may require the use of sprinkler heads that have higher operating temperatures. Table 6-2.5.1 in NFPA 13 (2013) contains seven temperature ranges and color-codings from Ordinary to Ultra High.

Figure 6-3 The red, liquid-filled frangible bulb on this upright sprinkler head, identifies it as an Ordinary (135°-170°) head. ©Depositphotos.com / Dave Willman

NFPA 13 attempts to minimize the amount of time that it takes for sprinkler systems to discharge water on fires by requiring that Ordinary temperature heads be installed unless the fire area meets an exception. (See figure 6-3.) This is important if fire spread is to be checked and early notification accomplished. In the design of sprinkler heads, **thermal lag** becomes the enemy.

NOTE

Sprinkler head spacing and discharge per square foot, known as density, are dependent on occupancy classification.

A sprinkler head that is rated to operate at 165°F does not operate the minute that the surrounding air reaches that temperature. The fusible link or frangible bulb must become heated to that temperature before it will operate.[3] In the case of a fire involving readily combustible material or flammable liquids, the surrounding temperature will have continued to rise and may be significantly higher by the time the first head operates and water is discharged.

The problem posed by thermal lag is minimized by designing sprinkler heads and other detection devices with less mass and greater surface area. This concept played an important role in the design of sprinklers for residential applications and is discussed with the other sprinkler standards.

NFPA 13 (see Table 6-2) uses an occupancy type system of hazard classification to determine the operational requirements of sprinkler systems, which is not to be confused with the use group classification from the building codes. Sprinkler head spacing and discharge per square foot, known as **density**, are based on occupancy classification.

Classification	Examples
Light Hazard	Churches, hospitals, schools, offices, theaters, residences
Ordinary Group 1	Parking garages, canneries, laundries, restaurant service areas, electronic plants
Ordinary Group 2	Dry cleaners, horse stables, machine shops, post offices, print shops, library stack rooms
Extra Hazard Group 1	Aircraft hangars, saw mills, plywood manufacturing
Extra Hazard Group 2	Flammable liquid spraying, plastic processing, varnish and paint dipping.

Table 6-2 Occupancy classifications from NFPA 13. Source: NFPA 13, 2013, Appendix A.

Using the occupancy classifications to design sprinkler systems for specific hazards, gives building owners an adequate system to protect their interest and that of their insurer, at a minimum cost. NFPA 13 systems provide excellent property protection by requiring that the entire building, including combustible void spaces such as attics, be protected.

A quick look at the examples in Table 6-2 may bring to mind some of the occupancies within your jurisdiction where the original owner has sold or leased the building to a business with a very different hazard classification. Ordinary Group 1 buildings are limited to storage heights of 8 feet, Ordinary Group 2 are limited to 12 feet. Determining that the hazard protected by the system has not changed is a key element in the inspections process, which we discuss in detail in Chapter 10.

Specific requirements for the design, installation, and acceptance testing of sprinkler systems are all contained within NFPA 13. Requirements for the maintenance and routine testing of sprinkler systems are contained in NFPA 25, *Standard for the Inspection, Testing and Maintenance of Water-Based Fire Protection Systems.* Sprinkler requirements for special hazards such as aerosol products, flammable and combustible liquids, nitrate film, and others are included in other NFPA standards. NFPA 13 Chapter 7 directs the user to the specific standards.

NFPA 13R Systems

NFPA 13R systems are designed to protect residential occupancies up to four stories in height. The systems enhance life safety by using listed residential sprinkler heads, which are designed to minimize thermal lag and respond quickly in the event of fire. By quickly suppressing the fire, **flashover** is prevented and occupants are able to escape from the building.

NOTE

By quickly suppressing fire, flashover is prevented and occupants are able to escape from the building.

NFPA 13R systems do not, however, provide the same level of property protection afforded by NFPA 13 systems. Combustible voids, as well as closets not over 24 square feet and bathrooms not over 55 square feet are not required to be sprinkler protected. The total building area exempted from sprinkler protection has been estimated as high as 67 percent.[4] The potential property damage resulting from fires that burn around, over and under a 13R sprinkler system is significant. The possibility of a total loss of a building equipped with a 13R system is very real. Hotels, motels, boarding houses, and apartment buildings are typically protected with NFPA 13R systems.

NOTE

NFPA 13R systems do not, however, provide the same level of property protection as afforded by NFPA 13 systems.

A trash fire on the balcony of the apartment building in Figure 6-3 quickly threatened the attic. Protected with an NFPA 13R sprinkler system, neither the balconies nor the attic spaces were sprinklered. Only quick action by the fire department saved the building.

Figure 6-3. This 3-story Type VB (combustible-unprotected) apartment required no rating for the 3rd floor roof-ceiling. The building was protected with an NFPA 13R sprinkler system, leaving the balconies without sprinkler coverage. Newlyweds moved into the apartment, unpacked and placed empty boxes and packing material on the balcony. A discarded cigarette ignited the combustible waste. An engine company returning to quarters spotted the fire, self-dispatched and was able to knock down the fire before the newlyweds were aware of the fire and before it entered the attic.

NFPA 13D Systems

NFPA 13D systems are designed to protect one- and two-family dwellings and manufactured homes. Like NFPA 13R systems, they are primarily designed to protect the occupants by preventing flashover from occurring and giving the occupants time to escape. They have similar limitations to NFPA 13R systems, in that bathrooms and closets as well as garages, attached porches, carports, and some foyers are not required to be protected with sprinklers. All sprinkler systems are not equal. They are designed for specific hazards and to perform specific functions.

Types of Sprinkler Systems

There are four basic types of sprinkler systems: wet pipe, dry pipe, preaction, and deluge. What differentiates them is the method in which water is supplied to the system. Each has specific applications and system-specific requirements from the sprinkler standards and in a few cases from the building codes themselves.

> **NOTE**
>
> **All sprinkler systems are not equal. They are designed for specific hazards and to perform specific functions.**

Wet Pipe Sprinkler Systems

Wet pipe sprinkler systems are by far the most commonly found systems in most jurisdictions. Water enters the sprinkler system from a dedicated supply or **fire line**, unless the system is a **limited area sprinkler** system or a system designed to protect one- and two-family dwellings and mobile homes. The supply valve is left open and water at street pressure is always on the system. For this reason, they are the quickest at getting water on the fire and are the simplest to maintain.

Wet pipe systems are installed where indoor temperatures can be maintained at or above 40°F. Below that temperature, there is the danger of freezing pipes. If you recall learning that water freezes at 32°F, your memory serves you well. If the outside temperature is below freezing and the interior temperature is less than 40°F, the steel sprinkler piping, which rapidly conducts heat and rapidly loses it, will drop below freezing. The frozen area may be isolated and near an opening or un-insulated portion of the building. It may be a small area, but it could be enough to put the whole system out of service.

NOTE

Wet pipe systems are installed where indoor temperatures can be maintained at or above 40°F, because below that temperature, there is the danger of freezing pipes.

Antifreeze systems are sometimes used where freezing is expected, but these systems are usually limited to 40 gallons or less. Each year, antifreeze systems must be drained into containers and the antifreeze solution must be replenished to restore the specific gravity required to prevent freezing. This process is costly and labor intensive. Larger systems subject to freezing are normally designed as dry pipe.

The wet pipe system's advantage of getting water to the fire area quickly is also its weakness in areas where water damage is of great concern. Mechanical damage from machinery such as forklifts or from vandals can lead to the soaking of valuable equipment, artifacts, merchandise, or records. The public's perception that all the sprinkler heads within a system discharge at once has caused undo concern on the part of many. Concern for items that may not be replaceable is legitimate and is easily addressed.

Dry Pipe Sprinkler Systems

Dry pipe sprinkler systems are installed in warehouses, parking garages, factories, and other buildings where there is a danger of freezing. They may also be installed in unheated attics or portions of buildings where providing heat is impractical. A **dry pipe valve** (see Figure 6-4) is installed at the sprinkler riser which keeps water out of the system piping until a fire activates a sprinkler head or heads. Dry pipe valves are designed so that a moderate amount of air pressure in the system above the valve (see Figure 6-5) is capable of holding back a much greater water pressure. The ratio of air pressure to water pressure at which the valve will open or trip is called the **differential**. Dry pipe systems are equipped with two pressure gauges. One gauge measures the incoming water pressure below the dry pipe valve, and the other measures the air pressure on the system above the valve.

Figure 6-4. Dry pipe valves are designed so a moderate amount of air pressure in the system above the valve, is capable of holding back much greater water pressure.

Figure 6-5. Top pane of photo shows a Viking dry pipe valve. Bottom pane shows the coverplate removed with the clapper valve in the open or "tripped" position. (Photos courtesy of David Feiring).

Air pressure for the system piping is usually supplied with a small compressor. The amount of air pressure required is dependent on the designed differential of the particular type of valve and the water supply pressure at the sprinkler riser. Differentials of 5 or 6 pounds of incoming water pressure to 1 pound of air pressure are common with traditional designs. Low differential valves, which operate with ratios of 1.0 and 1.2 pounds of water pressure to 1 pound of air pressure, have the advantage of up to a 70 percent reduction in the time required for water to flow from the system.[5]

NOTE

The delay between the point when the dry pipe valve trips and when water is discharged onto the fire is the fundamental weakness of these systems.

The fusing of a sprinkler head causes air pressure to drop within the system to a point where the dry pipe valve is tripped and latches open. Water then enters the system, displaces the air, and finally is discharged from the heads that have opened over the fire. The delay between the point when the dry pipe valve trips and when water is discharged is the fundamental weakness of these systems. This delay, which is effectively managed in most cases, makes the systems inappropriate for certain high hazard occupancies.[6]

In order to minimize this time lag, the sprinkler standards require dry pipe systems of more than 500 gallon capacity to be equipped with quick-opening devices,[7] unless water can be delivered to the most remote point of the system within 60 seconds. **Accelerators** and **exhausters** are designed to speed up the process.

The sprinkler riser and dry pipe valve must be installed within a heated enclosure and care must be taken to ensure that the system does not have low sections of piping where water can become trapped and not drained. Equal pressure readings on both gauges at the dry pipe valve indicate that the valve has tripped and the system is now wet. It will operate as a wet system but is subject to freezing. Inspection and maintenance is discussed in Chapter 10.

Preaction Sprinkler Systems

Preaction sprinkler systems are installed in properties where potential water damage from broken piping or sprinkler heads is of particular concern. Preaction systems are dry systems in which the water supply valve is opened on a signal from detection devices such as heat or smoke detectors. The potential for accidental water damage is minimized because no water is in the system unless a detection device has been activated or the valve is manually opened.

The sprinkler heads in a preaction system are traditional closed heads that must be fused by heat to open, discharging water only over the fire. Early notification also minimizes water damage associated with suppression of the fire because an alarm is transmitted upon activation of the detection device, which precedes the fusing of sprinkler heads and activation of a sprinkler flow switch.

Deluge Sprinkler Systems

Deluge sprinkler systems are installed in extra hazard occupancies where there is the possibility of a flash fire or fire growth so rapid that the response of a standard sprinkler system is too slow. Facilities with large quantities of flammable liquids or materials that pose a deflagration hazard are protected with these systems. The activation of the system is designed to apply water over the entire area covered by the system, rapidly and simultaneously. Deluge systems are dry systems that use open sprinkler heads and a deluge valve. The deluge valve operates much like the supply valve in a preaction system. When activated by a fire detection device, the deluge valve is automatically opened and water is discharged from all of the heads. Manual activation is also provided. Aircraft hangars, flammable liquid tank vehicle loading racks, and industrial facilities that process flammable or explosive materials are among those protected by these systems.

Acceptance Tests

Acceptance tests are required for all sprinkler systems and include a hydrostatic test and flush of the system piping, system operational tests, and main drain flow test. Hydrostatic testing of underground fire service mains (see Figure 6-6) should be conducted prior to backfilling. The witnessing of acceptance tests by code officials is time wisely invested. Table 6-3 summarizes acceptance testing requirements from NFPA 13.

NOTE

The witnessing of acceptance tests by code officials is time wisely invested.

Test	What/How	Why
Underground piping flush	Flow at 10 ft/sec through 4" min pipe or at max flow rate	Flush rocks and debris before connecting to the system
Hydrostatic test of piping	200 psi for 2 hours	Ensure system integrity under operating pressures
Dry pipe leakage test	40 psi for 24 hours	Ensure piping is capable of holding air under pressure
System operations tests	Trip test for dry pipe valves, pressure regulating devices, water flow devices and alarms	Ensure all valves, quick-opening devices, and alarms function
Main drain flow test	Open main drain, record static and residual pressures	Ensure all supply valves are open; used to compare with future tests to verify water supply

Table 6-3 Acceptance tests from NFPA 13.

ESFR Sprinklers

ESFR, or early suppression fast response sprinklers, were designed to protect palletized and rack storage up to 35 feet in some cases, with sprinklers mounted only at the ceiling level. ESFR systems use sprinkler heads with typical flow rates of up to 125 gpm per head, or about five times as much as a standard sprinkler. ESFR sprinklers are designed to extinguish fires. Standard sprinkler systems have always been designed to confine fires by pre-wetting combustibles that surround the fire.[8]

The technology is very attractive for storage occupancies. Storage racks can be reconfigured without sprinkler modifications, and storage heights are increased. As with any new technology, there are bugs to be worked out. Ductwork and lighting fixtures that might be installed by tenants after the sprinkler system is installed and approved can seriously compromise its effectiveness. Like any fire protection system, inspection to determine that the operational effectiveness of the system has not been reduced is of critical importance.

> **NOTE**
>
> **ESFR systems use sprinkler heads with typical flow rates of up to 125 gallons per minute per head, or about five times as much as a standard sprinkler.**

Water Mist Systems

Water mist systems suppress or extinguish fires by discharging a mix of water and nitrogen at more than 1,000 psi, out of the nozzle's tiny holes. The fire area is blanketed with a thick fog of water that extinguishes the fire by absorbing heat, displacing oxygen, and blocking radiant heat transfer.

Originally developed for enclosed compartments aboard ships, they were later used on passenger ships. The first land-based systems were installed in industrial structures by petroleum producers on Alaska's North Slope. Water mist systems are employed where weight (as in ships) or the risk of water damage justify the additional cost over traditional automatic sprinklers. Data centers, large commercial cooking operations, industrial and cooking facilities and cultural heritage properties such as Saint Patrick's Cathedral in New York and Thomas Jefferson's home, Monticello. [9]

STANDPIPES

Standpipe requirements are contained within the model building codes and in the referenced standard. Standpipe systems are classified according to their intended use, by the building occupants, fire department, or both. Table 6-4 summarizes standpipe supply requirements by type.

Table 6-4 Standpipe classes from NFPA 14

Class	Minimum Supply	Intended User
I	500 gpm for 30 minutes; 250 gpm for additional standpipes. Must maintain 65 psi residual @ 500 gpm.	Fire department
II	100 gpm for 30 minutes. Must maintain 65 psi residual @ 100 gpm	Trained fire brigades
III	Same as Class I. 2 ½ reducers installed to facilitate use by building occupants	Fire department and fire brigades

Source: NFPA 14, 2013, Section 3.3.17

Standpipes are generally required in high-rise buildings, covered mall buildings, buildings over three stories, large assembly occupancies, and adjacent to stages in theaters. Each of the model building codes has slightly different requirements. Generally, standpipe connections are required on each floor at every exit stairway, and on each side of a horizontal exit.

Standpipes must be capable of supplying 500 gpm at 100 psi (pounds per square inch) at the topmost outlet for the first standpipe, and 250 gpm for additional standpipes in the building. Previous editions of the legacy codes, predating the acceptance of the automatic nozzle by the fire service, required 65 psi.

Buildings under construction are generally required to have at least one standpipe or temporary standpipe, capable of flowing 500 gpm within one floor of the top of the building. These requirements are found within the respective fire prevention codes and are further discussed in Chapter 10, but are important to note here because they are not in all four building codes. You would not want to have to explain to the chief why the standpipe was not there.

Acceptance Tests for Standpipes

Acceptance tests for standpipes are similar to those for sprinkler systems and include the all-important flush, hydrostatic test, and flow test from the most remote discharge. A note here about **standpipe pressure-regulating devices**. As a fire pump pushes water up the standpipe, it must overcome the head pressure that the water column in the standpipe exerts at the base of the riser. It is .434 psi per foot or about 5 psi per floor. In order to pump water to the second floor and supply 100 psi of pressure at the standpipe, you would add an extra 5 psi or so. In order to get that same 100 psi at the standpipe on the twentieth floor you would have to add about 100 extra psi (20 X 5 psi). It gives you 100 psi on the twentieth floor, but 195 psi on the second floor (100 + (19 X 5)).

Standpipes are limited to 275 feet. Buildings that exceed 275 feet must have standpipes zoned to address this pressure problem, limiting the number of floors a standpipe can cover. Pressure-regulating devices are designed to control pressures within a single zone. The devices attach to the 2 ½-inch standpipe connections and restrict the opening, thereby limiting the available pressure. Two points crucial to the efficiency of the standpipe system as well as to the safety of the firefighters:

- The devices are designed for installation in a particular system and on a particular floor.
- They are preset at the factory and cannot be adjusted on the fire scene.

Figure 6-6 In order to deliver water at 100 psi. to the topmost outlet, the fire pump must overcome just under 5 pounds of head pressure per floor. Standpipe pressure reducing devices are installed on lower floors to moderate high pressure. Figure 6-6 demonstrates potential pressures without PRVs.

In our example of the twenty-story building in Figure 6-6, a pressure-regulating device on the second floor might reduce the opening of the orifice on the standpipe enough to decrease the available pressure by over 100 psi. What happens if that valve is mistakenly installed on the thirteenth floor? A crew operating off that standpipe connection might advance a hose line into a fire area and find that they cannot get water.

The problem came to the attention of the fire service when three Philadelphia firefighters were killed at the One Meridian Plaza fire.[10] PRVs were not a direct factor in their deaths. But they prevented effective attack lines from being deployed, hence prolonging the incident. If the buildings in your jurisdiction have not been inspected to determine that all pressure-regulating devices are installed on appropriate floors and will discharge correct flow and pressure, you are blindly waiting for a tragedy to occur.

Figure 6-7. Many standpipe pressure reducing valves are preset at the factory and are not field-adjustable. This model is field adjustable. Photo courtesy of A. Maurice Jones.

NOTE

If the buildings in your jurisdiction have not been inspected to determine that all pressure-regulating devices are installed on appropriate floors and will discharge correct flow and pressure, you are blindly waiting for a tragedy to occur.

Fire Department Connections

Fire department connections (FDCs) are required for all water-based fire extinguishing systems and standpipe systems with the exception of limited area sprinkler systems of less than twenty heads.

Fire department connections must be located on the street-side of buildings and be fully visible and recognizable from the street or nearest point of fire department vehicle access; or in a location approved by the chief.

Figure 6-8. The FDC must be located on the street-side of buildings and be fully visible and recognizable from the street or nearest point of fire department vehicle access; or in a location approved by the chief. Photo courtesy of A. Maurice Jones.

Private Hydrants and Fire Mains

Many people wrongly assume that all fire hydrants are part of the municipal water system and are installed and maintained as such. Hydrants are generally required wherever any portion of a building is more than 400 feet from an existing hydrant, and within 100 feet of a fire department connection to a standpipe system. This often necessitates the installation of private fire service mains and hydrants. (See Figure 6-9)

NFPA 24 *Installation of Private Fire Service Mains and their Appurtenances.* Fire flow, the required water supply at a site for use by the fire department for manual firefighting is included as IFC Appendix B, and UFC Appendix H.

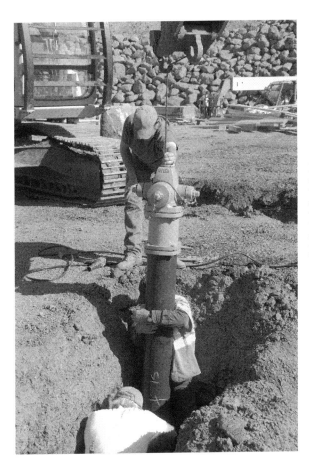

Figure 6-9 Private hydrant systems must be flushed and hydrostatically tested at 200 psi for two hours prior to connection with the municipal water supply. Source: NFPA 24, 2013 *Standard for the Installation of Private Fire Service Mains and their Appurtenances.* (Photo © Depositphotos.com/Tracy Fox).

Other Extinguishing Systems

Wet and Dry Chemical Extinguishing Systems

Wet and dry chemical extinguishing systems are used in various applications including range hood and duct fire protection (see Figure 6-10), paint spray booth protection, and even at unattended self-service motor vehicle fueling sites. Each has an NFPA standard that addresses the design, installation, and maintenance of the systems (NFPA 17A and 17, respectively). They use four significant mechanisms of extinguishing fires: smothering, cooling, radiation shielding, and chain breaking.[11]

Figure 6-10 Range hood fire suppression systems are designed and installed in accordance with NFPA 17-A.
(© Depositphotos.com/Sean Pryor)

The development of the UL-300 Standard for restaurant cooking areas by Underwriters' Laboratories generated considerable discussion within the fire community. The standard was adopted by UL in November of 1994 as a result of significant changes in the hazards posed in commercial cooking processes. First, cooking equipment has become more efficient. Heating elements are positioned on all four sides and on the bottom of fryers, not just on the bottom. Fryers are larger and better insulated.

> **NOTE**
>
> **In a commercial cooking operation, if the cooking equipment has not been upgraded and the frying medium is the same, then the hazard is the same and the system should be considered adequate.**

Where 50 pounds of oil was standard 30 years ago, 80 pounds is not unusual today, and in order to satisfy the health-conscious public, vegetable oils have replaced animal fats. Vegetable oils burn at higher temperatures than animal fats. If the quantity of overheated frying medium has doubled, then the heat absorption capability of the extinguishing system must also. Three to five times as much wet chemical agent as was previously required is required to meet the new UL-300 standard.[12]

Just as important for the fire inspector to realize is that this is not a retroactive requirement. Systems that have been previously approved (see Figure 6-8) have not been declassified by UL. The key issue is whether the hazard has changed. If the cooking equipment has not been upgraded and the frying medium is the same, then the hazard is the same and the system should be considered adequate. However, dry chemical range hood systems will gradually become extinct as the supply of replacement parts is exhausted.

Acceptance tests for wet and dry chemical extinguishing systems are prescribed by their respective standards and should always be witnessed by the code official.

Halogenated and Clean Agent Fire Extinguishing Systems

In addition to water mists systems, clean agent, halon, carbon dioxide (CO_2), and other extinguishing agents that do not leave a residue are the agent of choice for certain high-value commodities and equipment. During the late 1970s and early 1980s, halon was an acceptable topic of conversation at computer systems administrator's cocktail parties. Many seemed convinced that every sprinkler system was the deluge type, which in the event of fire would destroy their computer systems and lead to the financial ruin of their corporations. Many computer rooms were equipped with halon systems in lieu of sprinklers.

Halogenated extinguishing agents, or halon, are chemical compounds that contain carbon and one or more elements from the halogen series (fluorine, chlorine, bromine, and iodine). Most of the halogenated agents are toxic and unsuitable as extinguishing agents. Two compounds, Halon 1211 and Halon 1301, are effective extinguishing agents and are considered nontoxic.

Halon 1301 is used in total flooding systems, which discharge quantities sufficient to create up to a 10 percent concentration in the areas protected. Halogenated agents suppress fires by interrupting the chemical chain reaction, and do not extinguish deep-seated fire in class A materials.

Under the *Clean Air Act*, the United States government banned the production and importation of Halons 1211 and 1301 effective January 1, 1994, in compliance with the *Montreal Protocol on Substances That Deplete the Ozone Layer*. The ban raised questions regarding the continued viability of existing halogenated systems as well as halon fire extinguishers. The United States Environmental Protection Agency offers the following information at its Web site: www.epa.gov

- *Existing halon fire protection systems can continue in service. There are no laws which prohibit halon emissions, although EPA, NFPA and others all discourage the discharge testing of halon.*
- *Recycled halon can be purchased to recharge existing systems. The United States owns over 40 percent of the world's supply of halon 1301. The Halon Recycling Corporation or HRC, a nonprofit clearinghouse assists buyers and sellers.*

> **NOTE**
>
> **Existing halon fire protection systems can continue in service since there are no laws that prohibit halon emissions, although EPA, NFPA, and others all discourage the discharge testing of halon.**

In light of the government ban on halon manufacture and importation, other clean extinguishing agents such as Great Lakes Corporation's FM200, Dupont's FE-13, and Ansul's INERGEN have been developed. Fire protection for areas where the prevention of accidental water damage is a concern has also been rethought. The use of preaction sprinkler systems, which nearly eliminate the possibility of accidental discharge, has increased significantly, as has the popularity of water mist systems.

Carbon Dioxide Extinguishing Systems

Carbon dioxide is one of the most plentiful compounds on earth. It is a by-product of combustion and fermentation. We exhale carbon dioxide with every breath. It is cheap, leaves no residue, and is a highly effective extinguishing agent for class B and C fires. The major drawback to CO_2 systems is that carbon dioxide in concentrations of more than about 9 percent will render persons unconscious almost immediately,[13] so is not suited for the total flooding of computer rooms and other occupied areas. NFPA 12, *Carbon Dioxide Extinguishing Systems* contains installation, maintenance, and testing procedures.

> **NOTE**
>
> **The major drawback of CO_2 systems is that carbon dioxide in concentrations of more than about 9 percent will render persons unconscious almost immediately.**

Fixed Foam Systems and Water Spray Systems

Fixed foam and water spray systems are generally installed to protect hazards outside of buildings or for special hazards such as aircraft hangars. Foam systems protect flammable liquid storage facilities, and water spray systems are generally for exposure protection. Both have associated NFPA standards and specific acceptance tests that should always be witnessed by code officials, and also by the end user, the fire department. (See Figure 6-11)

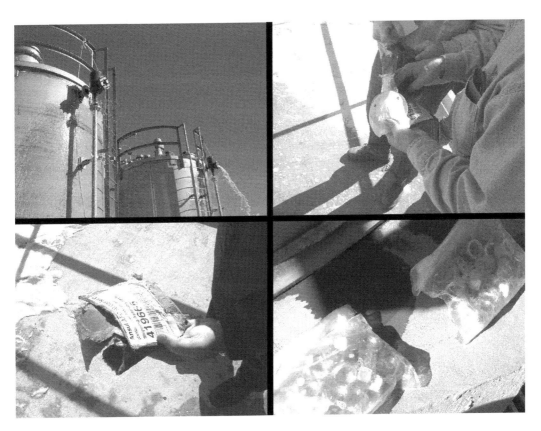

Figure 6-9 This photo collage is from an acceptance test for a foam system. The engineer of record swore the system had been pretested and another test was a waste of time. The fire official insisted and uncovered a serious problem with the system. (Courtesy of A. Maurice Jones).

Clockwise:

    A. Inadequate discharge during test
    B. Crew hands packing materials recovered from piping to the engineer.
    C. Packing material is identified as system specific
    D. Bag of *missing hardware* is recovered from piping.

Fire Extinguishers

Fire extinguishers are covered in depth in Chapter 10. They are nonetheless a vital element of a building's fire protection, that compliment built-in systems. Fire extinguishers are required to supplement some extinguishing systems such as range hood suppression and paint spray booth suppression systems. As with all other fire-extinguishing appliances and systems, the extinguishing agent must be compatible with the hazard and the other agents in use.

FIRE ALARM SYSTEMS

Fire alarm systems are required in certain buildings so that occupants receive prompt notification to evacuate, or in the case of institutional occupancies, for staff to take appropriate action. Fire alarm components are required in specific locations in all buildings. Smoke detectors are required in elevator lobbies to ensure that elevators do not return to the fire floor. Air handling units over 2,000 cubic feet per minute (cfm) must automatically shut down upon activation of a duct smoke detector installed upstream from the air handler. Smoke alarms are required in the sleeping areas of all residential occupancies and must be interconnected within individual dwelling units.

The design, installation, and maintenance of fire alarm systems are regulated by NFPA 72. In 1996, NFPA consolidated previous editions of NFPA 71, 72, 72E, 72G, 72H, and NFPA 74 into one document entitled NFPA 72, *The National Fire Alarm Code*.

Unlike most humans, fire alarm systems are quite good at doing several things at once. Fire alarm systems are comprised of a series of devices and circuits normally linked by a control panel. **Detection devices** identify the presence of heat or smoke, sprinkler flow, or manual activation by a building occupant, and signal the control panel. **Signaling devices** alert building occupants that there is a potential or actual fire in the building with audio and visual signals. Fire alarm systems also perform auxiliary functions such as elevator recall, automatic actuation of smoke removal or stairwell pressurization fans, and shutdown of certain air handlers. The system may also notify the fire department or a central station monitoring company.

Fire alarms range in size from the single station, battery-powered smoke alarm in an existing one-story home to a complex system in a high-rise building. Each has detection devices, signaling devices, and a control center. In the five-dollar battery-powered unit, they are all housed in a 6-inch plastic box. In a high-rise building, the circuitry may go for miles and cost many thousands of dollars.

Most complex modern systems are computerized. Fire alarm technicians do a significant amount of work within the software program during installation. When the flow switches are found to be annunciating in the fire control room as duct smoke detectors, technicians simply go into the program and make the correction with a few keystrokes. Before the advent of computerized systems, a technician with a screwdriver and pair of pliers had to move some wires to accomplish such a change—and that's a potential problem.

NFPA 72 requires acceptance testing of 100 percent of the system, and reacceptance testing any time components are added, deleted, or modified. This includes changes to software.[14] Additionally, reacceptance tests must include 10 percent of the initiating devices not affected by the modification. This requirement is necessitated by the complexity of modern systems and the sheer number of devices and functions involved.

NOTE

NFPA 72 requires acceptance testing of 100 percent of the system, and reacceptance testing any time components are added, deleted, or modified, including changes to software.

Schedules for routine testing and maintenance of systems as well as requirements for completion documents and for records of all tests are contained in NFPA 72. A review of these documents should be a routine part of the inspection of the building. Do not wait until after the fire when the system has malfunctioned to ask if the system was being maintained.

Emergency Alarm Systems

The IBC and IFC include requirements for emergency alarm systems for high-hazard occupancies. These are not fire alarms. They are systems that detect toxic and highly toxic gases, flammable gases, and refrigerants, and have manual pull stations that transmit a signal to a constantly attended location. They are people protection systems.

Smoke Control Systems and Smoke and Heat Vents

Systems that control the movement of smoke or provide for the rapid exhaust of smoke are required in atriums, covered malls, high-piled combustible storage facilities, underground structures, and large theaters. High-rise buildings and malls are required to provide manual controls for all air handling equipment so that firefighters can use the equipment for ventilation of the structure.

Smoke and heat vents are required by the model building codes in large factory and storage buildings where the length of exit access travel is long and over stages due to the large quantity of combustible sets. Curtain boards, which extend from the ceiling a minimum of 6 feet (but not within 8 feet of the floor), are installed to retard the lateral movement of smoke and gases. Smoke and heat vents must operate automatically and may be required to be provided with a means for manual activation by the fire department.

Summary

Fire protection systems are hazard-specific by design. The extinguishing method and capabilities of the system are designed to address a specific hazard. If that hazard changes, the system must be reevaluated.

All sprinklered buildings are not equally protected. NFPA 13, 13R, and 13D provide different levels of protection against different hazards. NFPA 13 provides protection to all areas of the building where combustibles (including building materials) may be present. In suppressing and perhaps extinguishing fires, they provide an excellent level of protection to the occupants of the building.

Residential sprinkler systems (NFPA 13R and 13D) provide a decreased level of property protection but optimize survivability of a fire by persons within the fire area by the quick response of the system. When installed in conjunction with smoke alarms, residential sprinkler systems provide an extremely high level of safety for building occupants from fires that originate within their dwelling unit.

The notion that fire resistance rated construction such as fire barriers and floor-ceiling assemblies can be totally replaced by sprinkler systems is a dangerous one. Fire walls between buildings provide property protection for each building owner. They are protected against accidents or the irresponsible actions of their neighbors by an assembly that requires no maintenance and is not easily breached. For a code official to trade that protection away on behalf of all the future owners of the property in return for a system that can be turned off or fail is questionable. Especially if the sprinkler system is designed for life safety, not property protection.

Thorough testing of fire protection systems, both during the approval or acceptance phase and for the life of the systems, is extremely important. The interests of the building owner as well as the contractor are best served by a formal program of 100 percent acceptance testing of all systems, witnessed by the fire official.

Review Questions

1. List the four classes of fire.

    a. _____
    b. _____
    c. _____
    d. _____

2. Fire is the rapid _____ of a material, accompanied by the evolution of heat and light.
3. What are the three legs of the fire triangle?

    a. _____

    b. _____

    c. _____

4. What are the four sides of the fire tetrahedron?

    a. _____

    b. _____

    c. _____

    d. _____

5. The difference between the operating temperature of a sprinkler head and the actual air temperature when the head operates is called _____.
6. Residential sprinkler systems installed in accordance with the NFPA 13R standard are designed to prevent _____ and allow occupants time to escape.
7. List three areas in which sprinkler heads may be omitted in buildings equipped with sprinkler systems designed under the NFPA 13R standard.

    a. _____

    b. _____

    c. _____

8. List the four basic types of sprinkler systems.

    a. _____

    b. _____

    c. _____

    d. _____

9. ESFR sprinkler system stands for _____.
10. List two changes in the commercial cooking process that led to the development of the UL-300 Standard.

    a. _____

    b. _____

Discussion Questions

1. A restaurant owner calls you to question the action of a fire equipment service company that attached a noncompliance tag to his range hood suppression system. They informed him that his system did not comply with the UL-300 standard and must be replaced. He informs you that since he fries very little, he is using the original cooking equipment from 1986, when the system was approved.

    a. Must he install a new extinguishing system?

    b. If not, in what circumstances would a new system be required?

2. You receive a call from the building inspector. He advises that a contractor has connected an underground fire main without flushing the system. The building is equipped with a standpipe and sprinkler system. The contractor proposes flushing the system with a 2½-inch hose line from the top standpipe outlet and flowing off the roof.

    a. Is this an acceptable method of flushing the system?

    b. How should the system have been flushed?

KEY TERMS

fire protection system

equipment and devices designed to detect a fire, sound an alarm and (or) make notification, control or remove smoke and hot gases, and control or extinguish the fire

fire

rapid oxidation accompanied by the evolution of heat and light

thermal lag

the difference between the operating temperature of a fire detection device such as a sprinkler head and the actual air temperature when the device activates

density

refers to sprinkler density calculated by gallons per minute discharge divided by the square footage covered

flashover

point at which the contents of a room or space becomes heated, simultaneously ignites, and the entire room or space becomes involved in fire

wet pipe sprinkler system

an automatic sprinkler system in which the supply valves are open and the system is charged with water under supply pressure at all times

fire line

dedicated underground supply piping for a sprinkler or standpipe system

limited area sprinkler system

an automatic sprinkler system that is limited to a single fire area and consists of not more than twenty sprinklers. The 2015 I-Codes lower the number of heads to five.

dry pipe sprinkler system

an automatic sprinkler system that features dry piping maintained under constant air pressure and a dry pipe valve in which water is held back by the pressure differential between the system air pressure and water supply pressure; used where there is a danger of freezing

dry pipe valve

sprinkler water supply valve designed to permit a moderate amount of air pressure above the valve to hold back a much greater water pressure from the incoming supply

differential

ratio of air pressure to water pressure that is necessary to balance a dry pipe valve, maintaining it in the closed position

accelerator

a quick opening device that permits system air pressure to enter the dry pipe valve below the one-way clapper valve, unbalancing the differential and causing the valve to trip more quickly

exhauster

a quick opening device used in dry pipe sprinkler systems that uses an auxiliary valve to discharge system air pressure to the atmosphere

preaction sprinkler system

an automatic sprinkler system that features dry piping, standard fusible heads, and a water supply control valve that is activated by fire detection devices; used to reduce the possibility of water damage from accidental breakage or discharge

deluge sprinkler system

an automatic sprinkler system that features open heads, dry piping, and a deluge valve that controls the supply of water, designed to wet the entire area upon activation

deluge valve

sprinkler water supply valve that is activated automatically or manually

early suppression fast response sprinklers (ESFR)

a type of fast response sprinkler designed to suppress fires in high challenge fire hazards through the application of increased flow densities

standpipe pressure-regulating device

a valve permanently attached to the standpipe discharge and designed to reduce flow pressure to a predetermined level by restricting the orifice size

halogenated extinguishing agent (halon)

a clean extinguishing agent composed of carbon and one or more elements from the halogen series (fluorine, chlorine, bromine, and iodine), which leaves no residue

detection device

a device connected to a fire alarm system having a sensor that responds to physical stimulus such as heat or smoke

signaling device

a notification appliance, being an alarm system component such as a bell, horn, speaker, light, or text display that provides audible, visible, or tactile output

Chapter 7

Means of Egress

Learning Objective

- Upon completion of this chapter, you should be able to:
- Identify and define the three components of a means of egress.
- Define the terms common path of travel, dead end, design occupant load, horizontal exit, and travel distance.
- Describe what is meant by the terms gross floor area and net floor area.
- Describe the six steps used in this chapter to design or assess the means of egress from a building or space.
- Calculate the required means of egress from a multipurpose space.
- Calculate the maximum occupant load for an assembly occupancy based on the exits provided.

INTRODUCTION: EGRESS AND THE CODES

The modern codes system employs a series of complimentary fire safety concepts. Building regulations are a bit of a balancing act designed to give us reasonably safe structures. All of the concepts incorporated in the codes are important, but if one had to be singled out for its role as the key element in many of our nation's disastrous fires, it would be means of egress. Often, other deficiencies such as combustible interior finish, high fire loads, or the improper use of open-flame devices are compounded by inadequate means of egress. In the end, the ability of the occupants to quickly and efficiently exit the building is often the difference between life and death.

> **NOTE**
>
> **All of the concepts incorporated within the code are important, but if one had to be singled out for its role as the key element in many of our nation's disastrous fires it would be means of egress.**

History is rife with catastrophic fires in which there was a large loss of life. The Iroquois Theater is just one of many fires where the large number of fatalities shocked the public. The Rhythm Club, Coconut Grove, Our Lady of Angels School, and Beverly Hills Supper Club conjure up stark black and white images of tragedies that plagued our past. To most of the public, they represent events from primitive times. *If they only knew then, what we know now.*

On February 20, 2003, foam-plastic soundproofing material was ignited by an indoor pyrotechnic display at the Station nightclub in West Warwick, Rhode Island. The fire claimed 100 lives and resulted in 200 injuries. Future generations will study the grisly images in color and on video rather than grainy black and white photographs. Perhaps they will think, *if they only knew then, what we know now.*

The Iroquois Theater fire, which occurred in Chicago in 1903, resulted in the deaths of more than six hundred people. It was later described as a "trifling fire"[1] that was quickly extinguished by the fire department, and was confined to the stage area and seats immediately in front of the proscenium. In the aftermath, bodies were reportedly piled three and four high in the aisles leading from the gallery.

Designed for 1,602 seats, over 1,770 were in place and nearly three hundred people were standing in the aisles behind the last row. The large volume of combustible theatrical sets and the lack of an appropriate fire separation between the stage and the audience, combined with overcrowding and exits that were obstructed or locked, created the worst theater disaster in our nation's history.

City ordinances that required skylights over the stage, an asbestos proscenium curtain designed to provide a separation between the stage and audience in the event of fire, flame-resistant scenery, and automatic sprinklers had simply not been enforced.[2] In the end, the "trifling" fire led to the deaths of more than six hundred people because they simply could not get out of the theater quickly enough.

In order to ensure that occupants are able to safely leave a building in any situation (not just fire emergencies), building and fire codes regulate construction features of the egress facilities and the occupant load. The codes:

- Establish a design occupant load based on the size and use of the space.
- Determine the number and capacity of exits based on the design occupant load.
- Establish a maximum travel distance from any point within the space to an exit.
- Require that multiple exits from a space be sufficiently remote from one another in case any one exit is compromised.
- Establish minimum ratings for corridors and exit enclosures.
- Establish criteria for direction of door swing, size, and methods of locking and latching.
- Require adequate lighting of all portions of the means of egress.

Elements of the Means of Egress

There are three distinct components of a means of egress: the **exit access**, the **exit**, and the **exit discharge**. The codes include the definitions within the code. Such a breakdown may seem trivial until we consider code provisions for maximum **travel distance**. Consider a modern high-rise building. The code is not regulating the distance from every point in the building to the lobby door at the sidewalk. It is regulating the distance from the most remote point on each floor to the enclosed, fire-rated stair tower. In a shopping mall, it's the distance from any point in the mall or business within the covered mall building to the fire-rated exit passageway that runs behind the stores.

In a high-rise building, the exit access begins at any point on the floor and terminates inside the rated stairwell. The exit begins inside the door of the rated stair tower and terminates outside the stairway discharge door at grade level. The exit discharge extends from the exterior door to the public sidewalk or public way.

> **NOTE**
>
> **There are three distinct components of a means of egress: the exit access, the exit, and the exit discharge.**

The only portion of the means of egress where distance is regulated is *exit access* (see Figure 7-1).

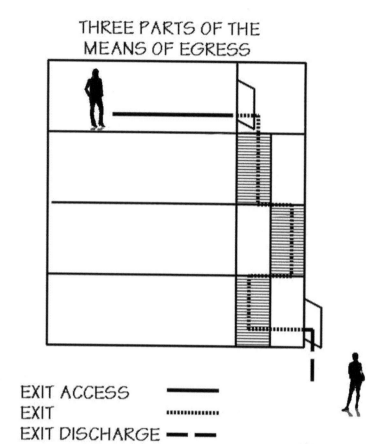

Figure 7-1. The three parts of the means of egress: Exit Access, Exit and Exit Discharge. Travel distance is regulated only is the Exit Access. Travel distance within an exit is theoretically, unlimited.

In essence, the code takes the viability of the exit for granted. It assumes that fire code provisions, prohibiting storage or any activity other than travel in an exit are vigorously enforced. It assumes all doors leading into the exit are maintained self-closing, and positively latching. This clearly demonstrates the necessity of an effective fire inspection program to ensure the reliability of the means of egress.

In small buildings, where the exterior door is within the maximum prescribed travel distance, exit access extends to the exterior door. The exit is only the thickness of the door itself, and the exit discharge terminates at the sidewalk or public way.

Code Provisions for Establishing Means of Egress

Throughout this text the model building codes are described as design documents. Design professionals are able to work through the provisions in a logical order to incorporate the code provisions within their plan for a building. The same steps are used by the code official or fire inspector in assessing the means of egress from a building or space, or determining the occupant load in a given situation, except in a different order.

Step 1. Calculating the Design Occupant Load

The first step in determining the required means of egress is to determine the number of persons who will occupy the space. Each of the model building codes has a similar table, which lists the use of the space and the square footage per occupant to be used (see Table 7-1).

It is important to note that the uses listed in the tables are not the use and occupancy classification of the building, but what the space is actually used for. The kitchen in a restaurant is not calculated as an assembly area, although the restaurant is a Group A-2 building.

Table 7-1. Maximum floor area allowances per occupant

FUNCTION OF SPACE	FLOOR AREA IN SQ. FT. PER OCCUPANT
Accessory storage areas, mechanical equipment room	300 gross
Agricultural building	300 gross
Aircraft hangars	500 gross
Airport terminal Baggage claim Baggage handling Concourse Waiting areas	20 gross 300 gross 100 gross 15 gross
Assembly Gaming floors (keno, slots, etc.)	11 gross
Assembly with fixed seats	See Section 1004.7
Assembly without fixed seats Concentrated (chairs only—not fixed) Standing space Unconcentrated (tables and chairs)	7 net 5 net 15 net
Bowling centers, allow 5 persons for each lane including 15 feet of runway, and for additional areas	7 net
Business areas	100 gross
Courtrooms—other than fixed seating areas	40 net
Day care	35 net
Dormitories	50 gross
Educational Classroom area Shops and other vocational room areas	20 net 50 net
Exercise rooms	50 gross
H-5 Fabrication and manufacturing areas	200 gross
Industrial areas	100 gross
Institutional areas Inpatient treatment areas Outpatient areas Sleeping areas	240 gross 100 gross 120 gross
Kitchens, commercial	200 gross
Library Reading rooms Stack area	50 net 100 gross
Locker rooms	50 gross
Mercantile Areas on other floors Basement and grade floor areas Storage, stock, shipping areas	60 gross 30 gross 300 gross
Parking garages	200 gross
Residential	200 gross
Skating rinks, swimming pools Rink and pool Decks	50 gross 15 gross
Stages and platforms	15 net
Warehouses	500 gross

Source: Excerpted from the 2012 *International Building Code;* Copyright 2011, Washington, D.C. International Code Council, Inc. Reproduced with permission. All rights reserved.

In using the design occupant load concept (see Figure 7-2) the code forces the designer to provide an adequate number of exits of the appropriate size for the number of people that should safely be able to use the space.

When calculating the design occupant load for assembly areas, **net floor area** is used. The net floor area is the actual square footage that people occupy. Deductions are made for stairways, columns and walls, restrooms, mechanical rooms, and other accessory areas, leaving only the area that is physically occupied. **Gross floor** area refers to the square footage contained within the outside walls including hallways and corridors, closets, restrooms, and other accessory spaces. Gross floor area is used for less hazardous spaces.

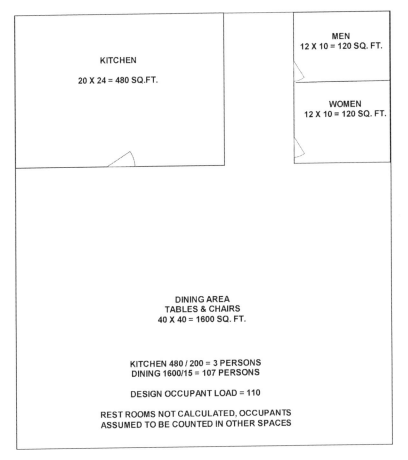

Figure 7-2. The design occupant load is calculated for every occupied space. Restrooms, coatrooms, and closets are not calculated as the occupants are assumed to have been included in other spaces.

Once the design occupant load has been calculated, the designer can determine the number, location, and capacity of the exits from the space. All the model codes contain provisions to have the occupant load and seating plan posted or available on the premises for assembly occupancies that have occupant loads exceeding fifty persons. Many occupancies contain multiple-use spaces, such as hotel ballrooms, where the occupant load for each use must be posted.

A typical ballroom will have two or more occupant loads posted-- one occupant load for functions with tables and chairs, such as banquets, one for chairs only, and one with no seating, where people stand. In our example in Figure 7-2, the floor area is divided by 15 square feet to determine the capacity of the dining room. In our ballroom example (see Figure 7-3), the occupant load for the concentrated assembly use is more than double that of the unconcentrated use. In this example, the ballroom would have been constructed with exit facilities for the largest of the design occupant loads. Not only must the doors leading from the ballroom accommodate the larger number, but the corridors and all other components of the means of egress must also.

Figure 7-3. Design occupant load is the theoretical number of people in a room or space, calculated by dividing the floor area by factors from the code. Exits must be adequate for the design occupant load(s).

Step 2. Determining the Number of Exits Required

With a few exceptions, every building and floor is required to be provided with at least two independent exits depending on the occupancy and occupant load. As shown in Table 7-2, all four model building codes for the most part, have identical requirements for the number of exits based on occupant load.

NOTE

In using the design occupant load concept, the code forces the designer to provide an adequate number of exits of the appropriate size for the number of people that should safely be able to use the space.

OCCUPANT LOAD (persons per story)	MINIMUM NUMBER OF EXITS (per story)
Based on occupancy, occupant load, travel distance & sprinkler system	1
1-500	2
501-1,000	3
More than 1,000	4

Table 7-2. Minimum number of exits. Low hazard buildings and spaces with low occupant loads are permitted to have only one exit based on travel distance, height about the level of exit discharge and sprinkler protection.
Source: Information excerpted from the 2012 International Building Code

Step 3. Determining Exit Remoteness

Having enough exits to accommodate the number of occupants of a building or floor area is fine, but what happens when the fire or emergency situation causes one of the exits to become blocked? The number of persons trying to use the remaining exit or exits is doubled or worse; what if the exits are so close that the same problem affects them both?

A fire that threatens the building next door becomes a whole new problem if the only exits from the exposed structure are on the same side of the building as the fire. Both exits could quickly become impassible. Even a relatively minor incident such as a vehicle fire could become a disaster.

The requirement for exit remoteness attempts to address the possibility that an exit may become blocked or impassible in an emergency situation. A secondary advantage of having exits remotely located is that occupants are spread throughout the space moving toward multiple exits and not congested in one area. In buildings or floor areas with two exits, the model codes require that they be separated by a distance, not less than one-half of the diagonal dimension of the room or space. In buildings and spaces with three or more exits, at least two must be separated by at least one-half the diagonal dimension, and the others arranged to minimize the possibility of an incident blocking multiple exits. (see Figure 7-4).

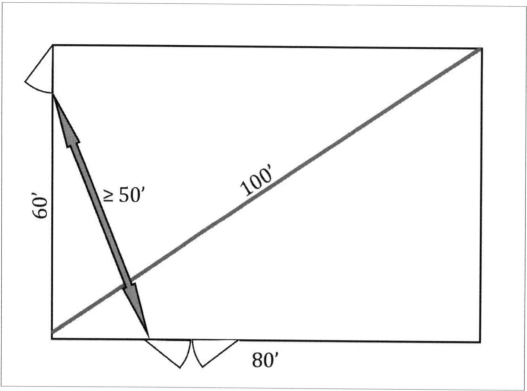

Figure 7-4. The exit access doorways must be at least 50 feet apart in this non-sprinklered building.

Note

The requirement for exit remoteness attempts to address the possibility that an exit may become blocked or impassible in an emergency situation.

In buildings that are equipped with an approved automatic sprinkler system, the distance may be reduced from one-half to one-third of the overall diagonal dimension.

Step 4. Limiting Maximum Travel Distance

We have established the number of exits needed for a particular building or floor area based on design occupant load and ensured that they are adequately remote to guard against a single fire or emergency situation blocking multiple exits based on the size of the room or space. We are now faced with the possibility that the exits are so far apart, or the floor area so large, that the distance the occupants must travel before reaching an exit is too great. We have described our building regulations as a balancing act. Limiting travel distance balances our desire to locate exits remotely with the need to enable persons to quickly reach an exit.

Note

Limiting travel distance to an exit balances our desire to locate exits remotely with the need to enable persons to quickly reach an exit.

The maximum distance that must be traversed before reaching a protected area such as an enclosed exit stair, exit passageway, exterior exit door, or horizontal exit or, in other words, the length of exit access travel is limited depending on the use group and whether the building is fully sprinklered.

The model building codes generally limit the length of **dead end** corridors or passageways to 20 feet. Twenty feet might not seem like much of a distance until you realize that for every step taken in a dead end, a return step must be taken after the occupants realize that they have gone the wrong way. There are exceptions, however, in sprinklered buildings and low hazard occupancies.

Dead ends commonly occur in buildings undergoing renovation with partial occupancy. An exit might become unusable due to construction and another provided. The area of the building nearest the decommissioned exit has the potential for a dead end situation to be created. It is important to consider this element when assessing the viability of the means of egress in a building undergoing renovation with partial occupancy or in buildings where illegal, non-permitted construction has taken place.

Figure 7-5. There are two exits from this floor, but occupants are essentially in a one exit space until they reach the corridor. This is the Common Path of Egress Travel.

Note

Dead ends commonly occur in buildings undergoing renovation with partial occupancy.

The model codes attempt to further protect building occupants by limiting the travel distance, before paths to two or more distinct exits become available (see Figure 7-5). The common path of egress travel is the portion of the exit access that must be traversed, before two distinct paths to two separate exits are available.

By regulating travel distance within the exit access, we are attempting to limit the amount of time that building occupants will be exposed to potential fire conditions. The IBC, NFPA 5000, and SBC take maximum travel distance a step further, by also limiting the travel distance before paths to two or more distinct exits become available. Regulating the **common path of travel** within the exit access limits the distance that building occupants must traverse before at least two distinct paths of travel to two or more exits become available.

Step 5. Calculating Egress Capacity

The only element lacking in establishing the exits from our building or floor area is how big we need to make them. How wide do the doors have to be? How about the corridors and stairs and the aisles that lead to them? All are addressed by the model codes, depending upon the design occupant load and use group. The design occupant load for the area served is multiplied by a factor to establish the width of the means of egress components.

The IBC and the legacy codes each have tables with the appropriate factors. The total occupant load multiplied by the factor provides the total capacity for the area served in inches, which is then divided among the required number of exits, but not without other specific requirements to keep in mind. The most obvious is a minimum size requirement for doors, ramps, corridors, and stairs, regardless of the occupant load served discussed in Step 6.

Exit Stairways Serving Multiple Floors

Where exit stairways serve multiple floors, the stairway capacity is determined using the floor with the largest occupant load, not with the occupant load of the entire building. The codes presume that in the time it takes for occupants to travel from the third floor landing to the second, the occupants of the second floor will have already made it to the first floor, and so on.

Where stairs converge, as in situations where exits from the basement and second floor meet and discharge on the first floor, the capacity of the exit from the first floor must be based on the occupant loads of the basement and the second floor combined.

Step 6. Ensuring Minimum Sizes for Means of Egress Components

For each element of the means of egress, there is a minimum size prescribed by the code, depending on the occupant load served (see Table 7-3). Regardless of the exit size based on capacity calculation, all components must conform with minimum size requirements. The net effect is that the minimum capacity calculated in inches is generally exceeded in order to conform with minimum size requirements. This is a plus for the safety of the occupants

	DOOR NET CLEAR OPENING	CORRIDORS SERVING ≤ 50	CORRIDORS SERVING > 50	STAIRS SERVING ≤50	STAIRS SERVING >50
IBC	32	36	44	36	44
BNBC	32	36	44	36	44
SBC	32	36	44	36	44
UBC	32	36	44	36	44
NFPA 101	32	36	44	36	44

Table 7-3. Minimum sizes for means of egress components

Corridors, Ramps, and Passageways

The required width of corridors ramps and passageways must be maintained to the public way. Projections such as drinking fountains or decorative trim are restricted depending on the model code, to a maximum of between 1 1/4 and 4 inches.

Doors that open into corridors are only permitted to reduce the corridor width by one-half and when in the fully open position, the door must be capable of folding against the corridor wall and projecting no more than 7 inches. Handrails are also limited to 3 1/2 inches on each side of the corridor, or 7 inches total.

Fire-resistance Ratings for Corridors

With some specific exceptions, corridors within occupancies that serve thirty people or more have to be rated for 1 hour. Those exceptions include fully sprinklered office buildings and a few others.

Smokeproof and Pressurized Enclosures

The travel distance to exit stairways within buildings is regulated, but once the occupants are within the exit enclosure they might have to descend three flights of stairs or thirty. The length of travel within the exit is not regulated. The model codes organizations recognized a need to ensure the viability of exits, particularly in high-rise buildings, where vertical shafts such as stairwells could act as chimneys for smoke and heat.

The exits from high-rise buildings (generally buildings with occupied floors more than 75 feet above the lowest level of fire department vehicle access) are given special attention by the codes. All exit stairs in high-rise buildings are required to be **smokeproof enclosures** or **pressurized enclosures**.

Smokeproof or pressurized enclosures are comprised of a 2-hour rated stairwell with rated opening protectives, connected to the floors of a high-rise by either exterior balconies that are open to the outside, or connected by ventilated vestibules. The stairshaft is also ventilated. The idea is to keep the smoke and heat out of the stairwell and vent it out the balcony or vestibule.

Main Exits from Assembly Buildings

Assembly buildings must have a main entrance and exit that discharges to the public way and is capable of accommodating one-half of the total occupant load. Additional exits must be provided to comply with requirements for exit remoteness, travel distance, exit capacity, and minimum exit size. The logic behind the requirement is that many of the occupants will immediately head for the main entrance in the event of an emergency. Although there are exits in every direction, many people will insist on going out the way they came in. They know that there is an exit there.

Note

Although there are exits in every direction, many people will insist on going out the way they came in.

Doors

With few exceptions, exit doors must be side swinging and open in the direction of egress when serving an occupant load of 50 or more or serving a high hazard occupancy. Panic hardware is required to release locks or latches on means of egress doors in assembly and educational occupancies, where occupant loads exceed 100, and in high hazard occupancies. Here we revisit the maintenance of rated assemblies discussed in Chapter 5.

Panic hardware that is installed on fire doors must be labeled as fire exit hardware. Door hardware that is labeled fire exit has been tested under fire conditions. Hardware that is labeled panic has been tested strictly for panic and not fire. It has not been designed to be part of a fire resistance rated assembly. Look for the testing laboratories mark on the hardware, or look it up in the appropriate directory.

Revolving Doors and Horizontal Sliding Doors

The death of 492 persons at the Coconut Grove fire in Boston in 1942 was due in large part to inadequate means of egress. Many exit doors were blocked or locked and the main exit, which most of the occupants were forced to attempt to use, was a revolving door.

There are significant restrictions regarding the installation of revolving doors. They cannot be installed within 10 feet of the top of a stair or escalator. Revolving doors must be capable of collapsing into a book-fold position that provides two parallel 18-inch wide exit paths. A conforming, side-swinging door must be installed within 10 feet, and in the same wall as a revolving door. In no case can revolving doors provide over 50 percent of the required egress capacity, and each revolving door is assumed to provide egress for no more than 50 persons.

Note

A conforming, side-swinging door must be installed within 10 feet, and in the same wall as a revolving door.

Figure 7-6. Revolving doors must have conforming side swinging doors within 10 feet. This modern hybrid locks in place on alarm and becomes two side-swinging doors leafs.

(© Depositphotos.com)

Many modern revolving doors (see Figure 7-6) are designed to lock in place on alarm and become twin side-swinging door leafs.

Horizontal sliding doors must be power operated with backup power supply, and be capable of being manually opened from both sides without special knowledge or effort in the event of power failure.

Door Width

Door width (see Figure 7-7) is determined by the actual clear opening provided and is not the physical measurement of the door. A 3'0" door (pronounced three-oh, for 3 feet 0 inches) is 36 inches wide when it comes from the factory. When it is installed in the jamb, the opening provided is reduced by the jamb itself, and sometimes by the hardware. That 3'0" door probably provides a 34-inch clear opening.

Figure 7-7. The actual clear width – not the size of the door leaf is used to calculate exit capacity.

The minimum clear opening is 32 inches. Often, multiple doors or leaves are installed at an opening. The maximum width of a swinging door leaf is 48 inches. When a door opening has two leaves, at least one of them must provide a clear opening of 32 inches. See Figure 7-8). When one leaf in a pair of doors does not provide 32 clear inches, it will not be equipped with hardware. This eliminates the possibility that occupants will mistakenly attempt to use the undersized door.

Figure 7-8. When one leaf in a pair of doors does not provide 32 clear inches, it will not be equipped with hardware.

Note

Door width is determined by the actual clear opening provided and is not the physical measurement of the door.

LOCKS AND LATCHES

The model building codes contain specific requirements for the locks and latches on egress doors. Unfortunately, these devices are often changed after tenant occupancy for security reasons, without regard to the original code requirements. The means of egress from a building or space may be totally adequate in location and capacity, but if the occupants cannot get out the doors because of an unapproved locking or latching arrangement, the exits might as well not be there. Exit doors must be openable from the egress side without the use of a key or special knowledge or effort, with certain specific exceptions.

Note

Exit doors must be openable from the egress side without the use of a key or special knowledge or effort.

Building and fire codes have traditionally required that the exit door must be openable from the egress side, without the use of a key or special knowledge or effort. Interior stairway doors are generally required to be openable from both sides, without the use of a key or special knowledge or effort. Security concerns over the years have resulted in code provisions that offer building owners the option to lock exits doors in some instances, provided additional safety features are provided. "Delayed egress locks," and "access-controlled egress doors" are two security arrangements permitted by the codes. They are drastically different systems with names that sound a lot alike. Inspectors should be constantly on the lookout for security locking arrangements that are installed without required permits that don't pass code muster and may pose a significant life hazard.

Business, factory, mercantile, and storage occupancies are permitted by all the codes to install a key-operated lock on the egress side of the main egress door, provided a sign is posted stating that the door is to remain unlocked when the building is occupied.

Locked Stairway Doors

Code provisions prohibiting locked stairway doors recognize the probability that stairwells can become untenable during fire conditions. Without the ability to reenter a floor and then exit by way of an alternate stairway, building occupants could become trapped. On October 17, 2003, a fire on the 12th floor of the 36-story Cook County Administration Building in Chicago resulted in eight fatalities and a dozen injuries. The dead were trapped in the stairwell by locks that had been installed on the stair doors. The City of Chicago's municipal code at the time was less stringent than the *Illinois State Fire Code*, which adopted the 2000 edition of the *Life Safety Code*.[3] *Life Safety Code* requirements for locked stairway doors included:

- Not less than two levels where it is possible to leave the stair enclosure to access another exit.
- Not more than four stories intervening between stories, where it is possible to leave the stair enclosure to access another exit.
- Reentry on the top story or next-to-top story allowing access to another exit.
- Doors allowing reentry identified as such on the stair side of the door.
- Doors not allowing reentry provided with a sign on the stair side indicating the location of the nearest door, in each direction of travel, that allows reentry or exit.

The *International Building Code* only permits locked stairway doors in high-rise buildings when they are capable of being unlocked simultaneously from the building's fire command center. In other than high-rise structures, doors can only be locked in stairs that connect not more than four stories.

Delayed Egress Locks

Delayed egress locks are installed on exit doors for security reasons. When the latching mechanism is activated, it sounds an alarm and remains latched for up to thirty seconds, allowing security personnel to respond before the door can be opened. The IBC prohibits them in assembly, education, and high-hazard occupancies. The *Life Safety Code* permits them on buildings with low and ordinary hazard contents. Both codes require the building to be protected with approved fire detection or suppression systems, and require that doors unlock upon activation of the fire protection system or loss of power to the locking mechanism. Both codes have additional requirements.

Access-Controlled Egress Doors

Access-controlled egress doors are locked from both sides but are equipped with sensors on the inside that automatically unlock and permit persons to exit. Each code has a laundry list of safety features that are required to ensure that building occupants can evacuate. Magnetic locks or "mag locks," are magnetic devices that work with no moving parts.

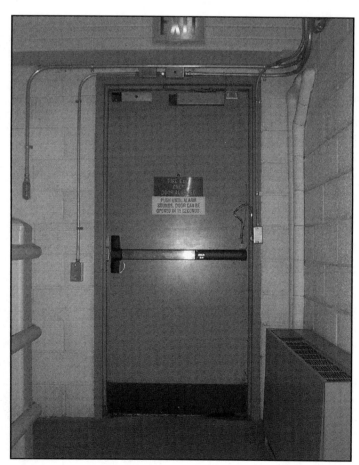

Figure 7-9. This rated stair-tower door is equipped with fire exit hardware and a magnetic lock (upper right). Activating the door's push bar activates an alarm and starts a timer to drop power to, and release the magnetic lock. (Courtesy of A. Maurice Jones).

Electric Strikes

Electric strikes or strike-plates are installed in place of traditional strike plates, for doors that are not required by code to remain latched when closed. Locksets are maintained in the locked position, prohibiting entry. When power to the electric strike plate is interrupted, it releases and pivots freely, allowing the door to be pushed open, even though the lock is still engaged and the bolt has not been retracted.

Electric strikes may be installed on security or perimeter doors that are not fire-resistance rated, provided they do not obstruct the means of egress. They may not be installed in rated stairway doors or doors in rated walls and partitions that must remain latched except when being opened.

EXIT ILLUMINATION AND EXIT LIGHTS

Except for dwelling units and inside of guest rooms in residential occupancies, the means of egress in all buildings must be provided with lighting with a minimum intensity of 1 footcandle at the floor. This includes the exit access, exit, and exit discharge. Theaters and auditoriums are permitted to reduce that level to 0.2 footcandles during performances.

Illuminated exit signs and emergency lighting of the means of egress are required, depending on the code, in occupancies that exceed certain occupant loads or are required to have more than one exit. Emergency lighting is only required to provide 1 footcandle for a specified period (1–1 ½ hours), and is designed to enable the occupants to safely exit the building, not to continue shopping or finish dinner. All the codes contain an exit sign exception for buildings where the main entrance serves as an exit and is clearly visible. In these buildings, no exit sign is required at the main entrance/exit.

ACCESSIBLE MEANS OF EGRESS AND AREAS OF REFUGE

Each of the model building codes has provisions that require buildings to be accessible to persons with disabilities. Specific provisions for **accessible means of egress** are provided for in the codes. The installation of these elements from the building code, are the responsibility of the building official. Responsibility for the maintenance of all means of egress, accessible or non-accessible, belongs to the fire official.

The requirements are new to some fire officials and introduce some totally new concepts in occupant protection. Unfamiliarity does not ease the burden of responsibility for enforcement of the maintenance provisions of the fire prevention code. New buildings are constructed with accessible means of egress complying with

ICC/ANSI A117.1, *American National Standard for Accessible and Usable Buildings and Facilities*.

Note

Responsibility for the maintenance of all means of egress, accessible or non-accessible, belongs to the fire official.

With few exceptions, all occupied spaces must provide an accessible means of egress. If more than one exit is required, at least two accessible means of egress must be provided. Each must be continuous, and include corridors, ramps, exit stairs, elevators that comply with specific requirements, horizontal exits, and smoke barriers. In order to be considered a part of an accessible means of egress:

Exit stairs must have a clear width of 48 inches between handrails unless the building is fully sprinklered. Elevators must be equipped with standby power. In nonsprinklered buildings, the elevator must be accessed through an area of refuge or horizontal exit.

Areas of refuge are spaces separated by 1-hour-rated smoke barriers. Doors must be rated for 20 minutes and equipped with approved hold-open devices that release upon smoke detector actuation. Each must be directly connected to an approved exit stairway or approved elevator and must have a two-way communications system.

ASSESSING THE MEANS OF EGRESS IN EXISTING SPACES

The same steps used in designing the means of egress for a building or space are used (in different order) to determine the exit capacity of an existing space. The concept of design occupant load, however, which we used to determine how many occupants our exits would have to accommodate, becomes the last step we consider, almost an afterthought. It is too late for that, the building is built. We are not as interested in how many theoretical bodies can fit into a given area as we are how many can get out-- how many bodies are the exits capable of accommodating.

In a perfect world, fire inspectors would never encounter a building without the approved occupant load posted on the wall. The mayor would not hold his reelection kickoff reception for three hundred guests (including the fire chief) in his campaign chairman's newly constructed strip warehouse space that lacks a final inspection. One thousand of the lovelorn would not show up for a book signing by a romance novelist at the local 2,500- square-foot bookstore. You would never be placed in a position to determine "just how dangerous is this?"

Step 1 – Determining the number of exits or exit access doors

The first step in determining the number of persons an existing space can accommodate is to count the exits. A 20,000 square foot warehouse with two exits cannot accommodate over 500 persons no matter the square footage and no matter how wide the doors are. Don't forget the requirement that at least two of the exits must be separated. Look at your code. The table below is a general synopsis:

PERSONS	EXITS
50 persons or less and meets travel distance and sprinkler requirements	1 exit
under 501 persons	2 exits
501- 1,000 persons	3 exit
over 1,000 persons	4 exits

Table 7-4. An approximate synopsis of single exit requirements across the codes.

Step 2 – Determining the capacity

Count the number of exit doors and determine the total net clear opening provided by them. The typical 3'0" door is going to provide about a 34-inch clear opening. Estimate the clear opening widths of the exit doors, total the figures, and then divide by the appropriate factor from the building code, generally .15" for sprinklered buildings and .2" for nonsprinklered. The resulting figure is the total number of occupants that the exit doors can accommodate. DON'T STOP HERE! (See Figure 7-10)

Step 3 – Ensuring continuity to the exit discharge

You have a nice large room with plenty of nice big doors. Let's assume they are adequately separated and swing the right way and are equipped with the proper hardware. The next step is to assess the means of egress components from the exit doors to the public way. Will the corridors accommodate the number of occupants that the doors did? Are there stairs? Measure the corridor and stair widths and divide by the appropriate factor in the building code. The resulting figure is the total number of occupants that each of the components can handle.

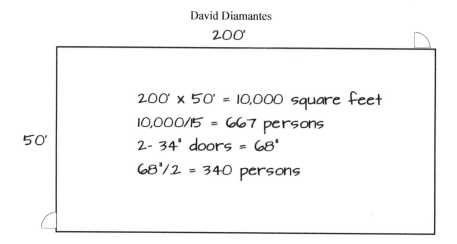

Figure 7-10. This 10,000 square foot nonsprinklered warehouse only has enough egress capacity for 340 persons regardless of the floor space and number of exits.

Are the doors side swinging, and do they open in the direction of egress? Is there appropriate hardware on the doors for the use and number of occupants? How about exit remoteness and travel distance? In most cases, you will be able to determine a safe occupant load without much of a problem. Usually you will be dealing with a single room or space.

WINDOWS FOR EMERGENCY ESCAPE AND RESCUE

Sleeping rooms in residential occupancies are required to provide at least one window for emergency escape and rescue unless they are sprinklered or have a door leading directly outside. Since 2000, windows for emergency escape and rescue have been required in all residential basements and a new feature has evolved—the climbable window well.

The installation of security bars or other locking arrangements is prohibited unless the window can be opened from the inside, without the use of tools or special knowledge or effort. A 2003 article in NFPA's Fire Journal attributed 19 fatalities and 26 injuries each year to "illegal gates and locks" that are installed in residential occupancies for security purposes.[4]

NOTE

Bars and unapproved security grills on emergency escape windows are a hazard to the public, and to firefighters.

Bars and unapproved security grills on emergency escape windows are a hazard to the public, and to firefighters. In 1993, seven children died in a Detroit house fire and eight members of a family were killed in Mississippi due to barred windows. In 1995, twelve people died in three separate incidents in California and four people in a single fire in Brooklyn, New York.[5] In each case the security devices prevented the occupants from escaping and proved to be an impediment to the fire department's rescue efforts. An internet search will uncover fatal fires almost every month in the U.S., most involving one or two victims, and unfortunately not rising to level to garner serious public attention.

NOTE

In order to install any system that affects egress from the structure, a building permit is required.

The installation of such devices is regulated by the building codes and the fire prevention codes. The obstruction of a required means of egress is a clear violation of the fire prevention code. In order to install any system that affects egress from the structure, a building permit is required. If the system requires the use of tools or special knowledge or effort, the installation violates the building code and should not be approved.

NOTE

A fundamental principal of exiting is that building occupants can never be exposed to a higher degree of danger when passing from one element of the means of egress to another.

Summary

The model codes recognize three distinct elements of the means of egress—the exit access, the exit, and the exit discharge. A fundamental principal of exiting is that building occupants can never be exposed to a higher degree of danger when passing from one element of the means of egress to another. The most hazardous area is the exit access. Occupants are exposed to the building processes, combustible storage and furnishings, and ignition sources. Travel distance within the exit access is regulated to limit our exposure in the event of fire.

Travel distance within the exit is not regulated. Absolutely no storage of any type is permitted nor is the exit to be used for any purpose other than the movement of people. The exit discharge is often overlooked by inspectors. The area between the termination of the exit and the public way must be maintained free of snow, storage, or any other material that impedes safe egress. All portions of the means of egress must be illuminated at all times.

Components of the means of egress are designed based on the occupant load of the building or space. The design occupant load is based on the use of the building or space. Illegal changes in use can have a severe impact on the ability of the occupants to safely exit a structure if the building was designed to accommodate fewer occupants.

Review Questions

1. List the three elements of the means of egress.

    a. _____
    b. _____
    c. _____

2. In which of the elements of the means of egress is travel distance regulated? _____

3. _____ refers to the square footage contained within the outside walls including, hallways, closets, and rest rooms.

4. In calculating _____, deductions are made for stairways, columns and walls, rest rooms, mechanical room, and other accessory areas.

5. Where exit stairs serve multiple floors, how is the stairway capacity determined? _____

6. Where exit stairs converge from the second floor and basement and discharge on the first floor, how is the capacity of the first floor exit determined?

    _____

7. Main exits from assembly buildings must be capable of accommodating _____ of the total occupant load.

8. All the model building codes require that a _____ be installed within 10 feet of every revolving door.

9. Door width is determined by measuring _____.

10. In what occupancies do the model building codes require windows as a part of the means of egress? _____

Discussion Questions

1. The owner of a local restaurant has installed an alarm lock device on his rear exit door. In order to exit, an occupant must first push or strike a red paddle clearly

marked "push to open," then open the door using the approved panic bar.

    a. Does this configuration meet the requirements of your building code?
    b. What locking options are available for the owner?

2. The IBC and the legacy model codes limit the distance the building occupants must traverse before entering a protected passageway or exiting the building.

    a. Why is the length of travel within the exit not regulated?
    b. Describe which building code provisions are designed to address this issue in covered mall buildings and high-rises.

KEY TERMS

exit access
that portion of a means of egress from any point in a building to an exit

exit
that portion of a means of egress that is separated from all other parts of a building by rated assemblies and provides a protected path to the exit discharge

exit discharge
that portion of a means of egress between the exit and the public way

travel distance
the length of the path a building occupant must travel before reaching an exterior door or an enclosed exit stairway, exit passageway, or horizontal exit; the total length of the exit access

net floor area
used in determining the design occupant load of spaces with high occupancy densities, such as assembly and educational areas; net floor area includes the area within the walls exclusive of the thickness of walls or columns or accessory spaces such as stairs, rest rooms, or mechanical rooms

dead end
a corridor, hallway, or passageway open to a corridor that can be entered from the exit access without passage through a door, but which does not lead to an exit

common path of travel
the portion of an exit access that building occupants must traverse before two distinct paths of travel to two exits are available

smokeproof or pressurized enclosure
an enclosed exit stair connected to all floors by either exterior balconies or ventilated vestibules and designed to limit the movement of smoke and fire gases into the stairwell

accessible means of egress
a means of egress including the exit access, exit, and exit discharge that can be entered and used by a person with a severe disability using a wheelchair and is also safe and usable for people with other disabilities

Chapter 8
Interior Finish Requirements

Learning Objectives

Upon completion of this chapter, you should be able to:
- Describe what building elements are regulated as interior finish.
- Describe the test method used to classify interior finish elements.
- Discuss the need to regulate interior finish based on previous fire experiences in the United States.

INTRODUCTION

There are several key elements that are common to fires where there is a large loss of life. Deficiencies in the means of egress, which are often compounded by overcrowding, delay evacuation, causing building occupants to be exposed to smoke and fire conditions. There have been many cases where occupants have inappropriately responded, underestimating the fire danger and not evacuating in a timely manner. Sometimes alcohol impairment is a factor. Highly combustible interior finish materials and decorations are the third common element. The speed at which a fire develops and spreads is highly influenced by the interior finish and types and amounts of combustible decorative material.

Highly combustible foam plastic material was installed in the Station nightclub in West Warwick, Rhode Island, after noise complaints about loud music. When the material was ignited by a pyrotechnic display, the fire spread rapidly reaching flashover in just over 1 1/2 minutes. The fire claimed one hundred lives and resulted in two hundred injuries. The speed at which the fire developed and the high percentage of fatalities are reminders of other fires where interior finish played a major role.

After the tragic 1942 Coconut Grove fire, in which 492 persons were killed and another 200 injured, a grand jury was convened. After 10 days of testimony by witnesses, the Suffolk County, Massachusetts, grand jury returned indictments against ten persons, including four officials from the city of Boston, the owner of the club, the wine steward, a contractor, a construction worker, and the interior decorator.[1]

Based on eye witness accounts that interior finish and decorative material played a significant role in the tragedy, Fire Commissioner William Reilly requested a chemist from the firm of Skinner and Sherman to test the material for flame resistance. The chemist's test report stated in part:

- The blue fabric from the molding in the Melody Lounge, where the fire reportedly started, burst into flame when touched with a match and was instantly consumed.
- Fibrous material wrapped around imitation palm trees burst violently into flames as a dry Christmas tree or excelsior would do.
- Straw and imitation straw matting did not ignite readily, but caught fire after a few seconds.
- The outer surface or coating of imitation leather wall covering from the lobby ignited readily and was consumed quickly. It was determined to be nitrocellulose.[2]

Table 8-1 lists fires in which interior finish played a significant role in a large loss of life and includes many names that are familiar to firefighters. Many of the incidents, such as the Coconut Grove, involved buildings with multiple deficiencies, where the lack of exits and highly combustible interior finish had a synergistic effect.

> **NOTE**
>
> **Many incidents resulting in a large loss of life involved buildings with multiple deficiencies, where the lack of exits and highly combustible interior finish had a synergistic effect.**

Fire	Location	Date	Fatalities
Rhythm Club	Natchez, MS	04/13/40	207
Coconut Grove	Boston, MA	11/28/42	492
LaSalle Hotel	Chicago, IL	6/05/46	61
Our Lady of Angels School	Chicago, IL	12/01/58	95
Beverly Hills Supper Club	Southgate, KY	5/28/77	165
Station nightclub	West Warwick, RI	2/20/03	100

Table 8-1 Fires where interior finish played a significant role.

Flame-resistance specifications for interior finish and decorative materials are spelled out in the model building codes. Provisions intended to maintain that level and to regulate decorative materials brought into the structure are included in the fire prevention codes.

Interior Finish and Trim

Interior finish and trim means the exposed interior surfaces of a building, meaning the ceiling, walls and partitions, columns, floor finishes, wainscoting, paneling, or any other finish applied structurally for insulation, fire resistance, acoustical, or decorative purposes. Interior finish contributes to fire development and intensity in four ways:

1. It affects the rate at which flashover occurs.

2. It contributes to fire extension.

3. It may intensify the fire by contributing fuel.

4. It may produce smoke and toxic gases.[3]

Interior finish materials other than floor finish must be classified in accordance with ASTM E84, Test Method for Surface Burning Characteristics of Building Materials; NFPA 255; and UL 723. All are designations of the **Steiner tunnel test**, developed by A. J. Steiner, an engineer at Underwriters' Laboratories.

In the Steiner tunnel test, materials are exposed to gas jets in a 25-foot horizontal furnace. Flame spread in the test material is then compared to asbestos cement board, which has an assigned flame spread rating of 0, and red oak, which has an assigned rating of 100. The higher the rating, the greater the potential hazard. Three physical characteristics are measured during the test: flame spread, smoke development or smoke density, and fuel contributed to the fire. As shown in Table 8-2, the BNBC and UBC used the numerals I through III, and the IBC, SBC, and NFPA 101 use the letters A through C to designate materials.

Flame Spread	IBC	BNBC	SBC	UBC	NFPA 101 & 5000
0–25	A	I	A	I	A
26–75	B	II	B	II	B
76–200	C	III	C	III	C

Table 8-2 Flame spread ratings

The model building codes regulate flame spread by specifying ratings according to use group classification, location within the means of egress, and sprinkler protection. In all cases, smoke development must be no greater than 450 for all materials.

Application

Interior finish applied to walls and ceilings must be applied directly to the structural elements or attached to furring strips. If furring strips are used, the void spaces created must be filled with inorganic or Class I materials, or fireblocked.

Carpet and Carpet-like Materials

Carpet and carpet-like materials that are installed as wall coverings pose a particular hazard. Less than 3 months after the disastrous MGM Grand Hotel fire, another Las Vegas hotel, the Las Vegas Hilton was struck by a fire that involved twenty-two floors of the thirty-story building. The fire was set with a small open flame device by a motel employee in an elevator lobby on the eighth floor.

The fire quickly involved the furnishings and the carpeting, which covered the walls and ceiling. After flashover occurred, the elevator lobby window failed, allowing a flame front to extend upward on the exterior of the building. Radiant heat exposure led to involvement and extension on upper floors, due in large part to the carpeted walls and ceilings of the elevator lobbies. Firefighters on the scene estimated that vertical extension to the top of the building took only 20–25 minutes.[4]

Carpeting that is manufactured as a floor covering is tested in a horizontal position, as it would normally be installed. The results of that test may have no relation to the actual burning characteristics of the carpet when installed on a wall vertically. All three model building codes require that carpet and carpet-like wall coverings shall either:

- Be Class I or Class A (0–25) and be installed in rooms or areas protected with automatic sprinklers or
- Meet the acceptance criteria when tested in accordance with NFPA 265, Standard Methods of Fire Tests for Evaluating Room Fire Growth Contribution of Textile Coverings on Full Height Panels and Walls.

NOTE

Carpeting manufactured as a floor covering is tested in a horizontal position, as it would normally be installed; the results of that test may have no relation to the actual burning characteristics of the carpet installed vertically on a wall.

Decorations and Trim

Decorations and trim are addressed by both building and fire codes. The building codes provide requirements to be addressed during design and construction. Fire codes provide requirements to be met by future tenants and occupants. Draperies, wall and ceiling hangings, and other decorations are regulated by occupancy type. The quantity of noncombustible decorative material is unlimited, but fire-treated material is limited to 10 percent of the total wall and ceiling area. The IFC and *Life Safety Code* contain provisions for the display of student-prepared artwork in schools.

Summary

The interior finish and trim and materials used for decorative purposes can pose significant hazards to building occupants. The model code regulations for these items were developed because of the severe fire potential of certain materials. Fires in which these materials contributed to rapid fire development and rapid fire spread have often resulted in numerous fatalities and injuries.

Review Questions

1. What is meant by interior finish? _____

2. List four ways in which interior finish contributes to fire development and intensity.
   a. _____
   b. _____
   c. _____
   d. _____
3. ASTM E84 and UL 723, are all designations of the _____ Test.
4. The test apparatus used in the ASTM E84 and UL 723 is a furnace that tests materials in a _____ position.
5. List the three characteristics that are measured during the ASTM E84 and UL 723 tests.
   a. _____
   b. _____
   c. _____
6. According to your model building code, what is the classification of a material that has a flame-spread rating of 10? _____
7. In a sprinklered building, what is the minimum flame-spread rating for carpeting that is to be installed as a wall covering? _____
8. What is the maximum smoke development permitted for interior finish materials? _____
9. Name two major fires in which interior finish was a significant factor.
   a. _____
   b. _____
10. Is exposed insulation within an occupied space considered interior finish? _____

Discussion Question

1. You receive a call from the chief. He has been approached by the local church whose request to install carpet on the walls of their non-sprinklered sanctuary was denied by the building official, based on your recommendation. The chief voices the same concern as the church board of directors: "It is Class I, the safest possible rating."
   a. Explain the basis for your decision to the chief.
   b. Would the installation of smoke detectors in the sanctuary be a satisfactory solution?

KEY TERM

Steiner tunnel test

test to determine the surface burning characteristics of building materials in which the flame spread of the test material is compared to asbestos cement board, rated 0, and red oak, rated 100; the higher the rating, the greater the potential hazard.

Section 3

The Fire Code

In the introductions to Sections 1 and 2, we discussed the importance of the roles and responsibilities of code officials and enforcing agencies. Hybrids, in which multiple disciplines are the responsibility of a single agency, have become more common as governments reorganize and consolidate. More and more code officials wear several hats and enforce several codes, a situation that forces all of us to constantly bear in mind our legal responsibilities, duties, and authority.

For the remainder of the text, get out your fire official hat. We will discuss the safeguarding of life and property from fire and explosion arising from the storage, handling, and use of hazardous substances, materials, and devices, and from conditions hazardous to life and property in the use or occupancy of buildings or premises, through the use of the technical provisions of the fire code. You will see very quickly that the fire code does a lot more than just keep fires from occurring. It contains provisions for minimizing the impact of the fires that occur despite our best efforts at prevention.

In the first three editions of this text, I referred to the "fire prevention code." Two of the legacy codes, Southern Buildings Code Congress, International's, *Standard Fire Prevention Code* and Buildings Official's and Code Administrators, International's, *BOCA National Fire Prevention Code,* contained the word *prevention* in their titles. The *Uniform Fire Code,* maintained by the Western Fire Chiefs Association and currently published as NFPA 1, left out the word "prevention," because the code is much more than a document aimed at preventing fires. In 2000, the International Code Council followed suit when they released the first *International Fire Code*. In this 4th edition, I've made every effort to get on board and refer to the documents as *fire codes*; but old habits die hard. I beg your indulgence ahead of time, if my edits weren't uniform.

Chapter 9
General Fire Safety Provisions

Learning Objectives

Upon completion of this chapter, you should be able to:
- Describe NFPA 550, The Fire Safety Concepts Tree and its uses.
- List the two strategies for fire safety used by the model fire prevention codes.
- List three objectives for achieving the goal of life and property protection using each strategy.
- Describe the significance of illegal changes in use and list the code sections that apply.
- List four rated assemblies that an inspector would routinely check to ensure compliance with the code.

Introduction: General Provisions for All Occupancies

Successful fire prevention is a bit like football. You can have all the razzle-dazzle plays in the world, but if your team cannot block on the line and cannot tackle, you will not win. These are not recent revelations. Lombardi said it, Landry said it, and every successful NFL coach is still saying it. Forget the basic fundamentals, and no matter how complex your system, you will not overcome your basic weaknesses.

Despite all the technological breakthroughs that are occurring at an accelerating rate, the fundamentals of fire prevention and protection have not changed since a.d. 64, when Emperor Nero was said to have fiddled while Rome burned. We have not come up with anything yet to replace what we have learned the hard way over and over since the very beginning of our civilization: (1) eliminate ignition sources, (2) control the available fuel, and (3) prepare for the times you fail at numbers 1 and 2.

> **NOTE**
>
> **The fundamentals of fire protection and prevention have not changed since A.D. 64, when Emperor Nero was said to have fiddled while Rome burned.**

In 1974, NFPA published the first edition of NFPA 550, *Guide to Fire Safety Concepts Tree*. The guide identifies fundamental strategies and lists objectives for achieving a fire safe environment. The guide is extensive and is well worth studying it. The fault tree analysis system was first developed by H. A. Watson of Bell Telephone labs in 1962 and made famous by Boeing Corporation in its application to the Minuteman ballistic missile program.[1]

Figure 9-1 presents the relationship between fire and building codes also in a fault-tree analysis format, yet different then NFPA 550. The fire codes attempt to prevent fires while the fire protection features of the building codes are geared at managing the impact of a fire. The relationship provides valuable insight for the inspector whose job is to ensure the viability of those design features for the life of a building.

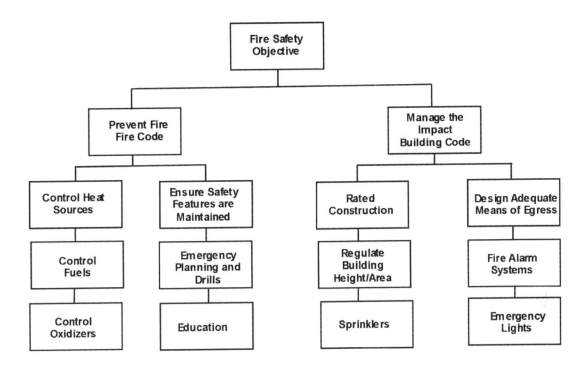

Figure 9-1 represents the relationship between fire and building codes.

The goal of the model fire and building codes, which is the safeguarding of life and property is achieved through two basic strategies: to prevent fires from occurring, and, when they do occur through some failure, to manage the impact on people and property.

Consider the building code fundamentals discussed in Chapters 3 through 8. Regulating use and occupancy, requiring fire-resistive construction elements and built-in fire protection systems, regulating interior finish, and requiring adequate egress facilities are all functions of the model building codes. These features prescribed by the model building codes do not keep fires from starting. They are aimed at minimizing the impact of fire.

Fire codes contain provisions to keep hostile fires from occurring, as well as those that attempt to minimize the impact of fires by requiring the maintenance of the fire safety construction features required by the building codes.

Prevent Ignition

Prevent the ignition of fires by effectively segregating the three elements of the fire triangle.

- Control heat sources such as smoking, unsafe heating appliances, electrical hazards, open flames, and burning.
- Control fuels by regulating the storage and handling of specific materials.
- Control the interaction of fuel, oxygen, and the heat source in storage, handling, transportation, and processing.

NOTE

Prevent the ignition of fires by effectively segregating the three elements of the fire triangle.

Manage Impact

- Manage the fire impact by effectively regulating construction, use and occupancy, and occupant training.
- Limit building height and area based on the combustibility and fire resistance of construction elements.
- Provide adequate fire resistance for structural elements and separation assemblies.
- Provide adequate emergency egress facilities.
- Provide built-in fire protection systems to detect and suppress fires.
- Ensure that hazardous processes and occupancies are adequately separated from other occupancies.
- Ensure that occupants and employees in critical occupancies receive appropriate training in evacuation and emergency response.

NOTE

Manage the fire impact by effectively regulating construction, use and occupancy, and occupant training.

As we study the provisions of the fire codes, trace each one back to the basic fire prevention fundamentals. These same fundamentals were behind the development of most non-model code-related fire prevention ordinances adopted by state and local governments over the years.

CHANGES IN USE AND OCCUPANCY

Of all the fire code violations that you will ever encounter, the illegal change of use presents the largest potential for the number of individual safety deficiencies that can exist. As we learned in Section 2, each use group has specific requirements that range from limits on building height and area and the installation of fixed fire protection to the establishment of means of egress. An illegal change in use can take a building that conforms to the requirements of the original use group and create a situation in which the means of egress and fire protection features are totally inadequate, often to the surprise of the occupants.

A 15,000-square-foot Business Use Group building, that is illegally converted into a house of worship easily results in an occupant load that is more than fourteen times greater than that which was used to design the exits for the original Use Group B building! The original use did not require a sprinkler system, fire alarm system, or panic hardware on exit doors. Now more than 2,000 people have congregated in a building with only enough exits to quickly evacuate about 150 people, and without the fire protection features that years of experience have proved are vital to prevent catastrophes.

What is the violation? Should the inspector cite a violation of Chapter 10, *Means of Egress*, of the building code? How about Chapter 9, *Fire Protection Systems*?

The violation that the fire code official should cite in this instance is "illegal change in use." The lack of adequate means of egress or inadequate fixed fire protection systems may be the overriding concern, but the actual violation is the illegal change in use. Lack of sprinklers, adequate exits, and panic hardware are the symptoms; illegal change in use is the disease.

Don't forget that a change in use necessitates compliance with building, mechanical, plumbing, and electrical codes, not to mention the jurisdiction's zoning and other regulations. In some instances, changes in use are impossible without literally demolishing the building and rebuilding with different construction materials.

Course of Action

The prudent inspector takes the following actions when illegal changes in use are identified: (1) take immediate steps to ensure the safety of the occupants, including issuing and emergency evacuation order, if warranted, (2) issue a Notice of Violation prohibiting the illegal use or activity, and (3) report the condition to the appropriate code officials and coordinate your follow-up activities with them. Simply "turning it over" to another official is not adequate. You need to follow through.

Issuing a violation under the building code will get your notice overturned in court or before the appeals board, unless you are with one of the fire prevention bureaus specifically empowered to enforce the building code. Remember what hat you are wearing!

Other illegal changes in use are more subtle, such as a change from low to moderate hazard storage, or from one assembly sub group to another. The quantities of regulated materials stored or handled within a building can also create an illegal change in use. The terms *regulated* and *hazardous* materials, as used within the model code process, are not limited to the chemicals that only the members of the hazmat team can spell. Hazardous materials include physical and health hazards regulated by the codes, and run the gamut from combustible liquids to ammonium nitrate fertilizer.

Other regulated materials include combustible fibers, combustible metals, and even baled wastepaper. These are the same items that fire codes regulated before the terms *hazardous materials* and *hazmat* became popular. The storage or use of these substances in quantities that exceed a prescribed threshold, or *maximum allowable quantity*, constitutes an illegal change in use. There is more on hazardous materials code requirements in Chapter 12.

NOTE

Hazardous materials include physical and health hazards regulated by the codes, and run the gamut from combustible liquids to baled wastepaper and ammonium nitrate fertilizer.

One of the greatest challenges facing fire and building officials as well as fire departments, is the growing number of senior citizens who need to live in a structured environment, yet fall somewhere short of being "incapable of self-preservation," and requiring institutional care. Structures that do not have the built-in fire protection features required for hospitals and nursing homes are being used more and more to house elderly persons, who may or may not be capable of appropriate response and evacuation in the event of a fire emergency. Political expediency, political correctness, and short-sighted economics all tend to point toward lax enforcement of the fire code in this instance.

Unwillingness on the part of some fire officials to rectify illegal changes in use when facilities are used for nonambulatory residents and do not meet the strict requirements for institutional occupancies, will inevitably lead to fires with injuries and deaths.

The use group classification system of the *BOCA National Building Code* was the subject of a lawsuit brought under the *Americans with Disabilities Act* and Federal *Fair Housing Act* in the federal court system in 1995.[2] At issue, was the building code use of disability (capacity for self-preservation) as a means to determine occupancy classification. The case was dismissed, but others will inevitably follow. Is it better to be sued for a good faith effort at enforcing a legally adopted, nationally recognized model code; or is it better to be charged for criminal nonfeasance and be sued after a disastrous fire for not enforcing it?

Fire Prevention, Inspection and Code Enforcement

MAINTENANCE OF RATED ASSEMBLIES

The model building codes take a few things for granted. Some of them, like gravity, are sure bets. Buildings try to fall down, not up. Others, like the continued viability of fire barriers, rated corridors (see Figure 9-2), and exit enclosures are not such good bets, unless there is an ongoing, aggressive inspection program that ensures their survival. Simply put, all required fire-resistance rated assemblies and required barriers to fire must be maintained at all times. This list includes walls, partitions, floor/ceiling assemblies, stairway enclosures, draftstopping, fireblocking, firestopping, column protection, and all other rated assemblies and all opening protectives.

NOTE

Simply put, all required fire-resistance rated assemblies and required barriers to fire must be maintained at all times.

Figure 9-2. This rated exit passageway has been compromised and the walkway obstructed during construction and poses a potential threat to the occupants of the building. Such work should be performed and completed after hours.

A primary consideration during every inspection must be whether required ratings have been violated by illegal construction or the addition of improper equipment or hardware. This determination is not possible if the inspector does not know whether rated assemblies were required in the first place, so the building code must always be within arms' reach.

> **NOTE**
>
> **A primary consideration during every inspection must be whether required ratings have been violated by illegal construction or the addition of improper equipment or hardware.**

Rated Walls, Partitions, and Floor/Ceiling Assemblies

Rated walls, partitions, and floor/ceiling assemblies can be illegally penetrated in order to combine two tenant spaces, run utilities such as communications cables, or for a host of other reasons. The code is explicit that repairs must restore the assembly to its original rating through an **approved** method. A cosmetic sheetrock repair, in which a partition is merely patched and may look fixed, is not sufficient.

Inspections of corridor walls and tenant separation walls often require a look above the ceiling to determine that walls extend to the roof deck. If the area above the ceiling is used as a return air plenum and the building is of noncombustible construction, this is the time to check that any cable installed is listed or labeled as being approved for plenum use.

> **NOTE**
>
> **Repairs must restore the assembly to its original rating through an approved method; a cosmetic sheetrock repair, in which a partition is merely patched and may look fixed, is not sufficient.**

Structural Elements

Ratings for structural elements such as columns and beams are particularly prone to damage in warehouses and buildings where occupants have access and are forced to work around the assemblies. Employees operating forklifts or other machinery can damage protective assemblies of gypsum or spray-applied fire-resistant material. You can order the repair of the assembly at every inspection, and come back a week later to find that it has been damaged again. Impact protection, with metal corner guards or a substantial jacket of metal or other noncombustible material should be required and considered an integral part of the rated assembly.

Wires, cables, and pipes are not permitted to be embedded in the fire protective covering of a structural member that is individually encased. Likewise, gypsum or spray-applied fire-resistant material cannot be penetrated by tradesmen attempting to use the protected structural element for some purpose.

> **NOTE**
>
> **Impact protection, with metal corner guards or a substantial jacket of metal or other noncombustible material should be required and considered an integral part of the rated assembly.**

Draftstopping and Fireblocking

Draftstopping and fireblocking are barriers to limit fire spread that are not rated assemblies. The building code requirements for these two components are prescriptive and are not based on performance. Most fireblocking will be hidden from view behind wall and ceiling construction. Draftstopping, especially in apartment buildings of combustible construction, is a critical element that should be checked during the inspection. Although draftstops are not rated assemblies, the model building codes require that their "integrity be maintained."

Builders have the option of providing an approved access opening to each draftstopped area or providing openings within the draftstopping. Openings in the draftstops must be self-closing and positively latching. Access openings in rated ceilings must also be protected with listed opening protectives.

> **NOTE**
>
> **Although draftstops are not rated assemblies, the model building codes require that their "integrity be maintained."**

Opening Protectives

Opening protectives such as fire doors, fire windows, and fire dampers are required to be accessible for inspection and labeled at the factory with a label or listing mark indicating the fire protection rating. If the label or mark is not there, it is not a rated *assembly. The label is a required component of the assembly, just as the latching* mechanism of a fire door is a required component. If there is no label on the opening protective, it must be replaced.

NOTE

The label is a required component of the assembly, just as the latching mechanism of a fire door is a required component.

Side-swinging fire doors and smoke doors must be self-closing and positively latching. Doors must close and latch when released from any position, from fully opened to nearly closed. The installation of hardware such as special locks, mail slots, and door alarms are prohibited unless the hardware is listed for installation in the door. NFPA 80: *Fire Doors and Windows* contains specific provisions regarding the modification of fire doors.

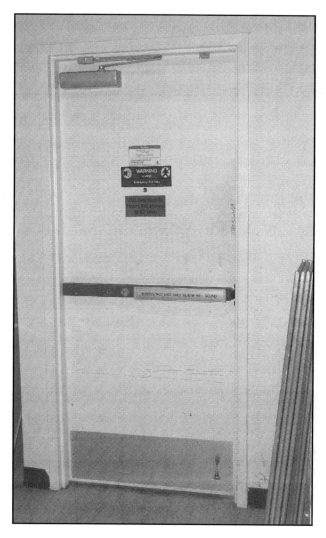

Door stops, wedges, and other hold-open devices are prohibited on fire doors (see Figure 9-3). Approved hold-open devices that are self-closing upon the activation of automatic fire detectors are permissible, but must be installed under a building permit.

Figure 9-3. Door stops, wedges, and other hold-open devices are prohibited on fire doors.

Rolling fire doors (see Figure 9-4) must be tested annually by the building owner and records maintained on the premises. If you do not ask for the records during your inspection, you can count on two things: (1) they are not keeping the records and (2) they are also not testing the doors. If you wait until the fire occurs in which something awful happens and then ask for the records, you are too late. Fusible links that operate rolling fire doors must be kept clean and unpainted.

Figure 9-4. Rolling fire doors must be tested annually by the building owner.

DECORATIVE MATERIALS AND INTERIOR FINISH

In Chapter 8, we discussed the role that combustible interior finish materials have played in major fires and how the model building codes were changed to address the problem. Although it is closely regulated by the model building codes, interior finish continues to play a significant role in fatal fires.

Decorative wall coverings are often installed without the benefit of the building permit process, either through ignorance or by unscrupulous contractors who choose to ignore *the regulations. Carpetlike textile wall coverings, wood paneling, and even plastic* decorative panels are installed over noncombustible gypsum wallboard. This significantly adds to the fuel load, increases potential fire spread, and accelerates flashover. Although interior finish materials are generally not the first material ignited in a building fire, wall and ceiling finish plays a major role in fire development and fire spread.[3] In the 2003 fire at the Station nightclub, foam plastic that had been applied to the walls for acoustics' purposes was the first material ignited, and was perhaps the single most significant factor in the large loss of life at the West Warwick, Rhode Island nightclub.

NOTE

Although interior finish materials are generally not the first material ignited in a building fire, wall and ceiling finish plays a major role in fire development and fire spread.

After a 1990 fraternity house fire in Chapel Hill, North Carolina, NFPA's senior fire investigator identified the building's 1 inch by 8 inch floor-to-ceiling, southern pine paneling as a major culprit in the five deaths that occurred.[4] The paneling had been installed on furring strips that had been nailed to the building's plaster walls. "If it hadn't been for the wood paneling, which created a very intense fire in the basement, this tragedy might not have happened."[5] The IFC requirements for interior finish in existing buildings are shown in Table 9-1.

Decorative materials brought into buildings for holiday displays, such as cornstalks, straw bales, and Christmas trees pose a safety problem as well as a potential public relations nightmare without good communications with business owners.

Figure 9-5. A fire involving the interior finish or decorative material in a covered mall has the potential of affecting the entire covered mall building.

GROUP	SPRINKLERED			NONSPRINKLERED		
	Interior exit stairways and interior exit ramps and exit passageways[a,b]	Corridors and enclosure for exit access stairways and exit access ramps	Rooms and enclosed spaces[c]	Interior exit stairways and interior exit ramps and exit passageways[a,b]	Corridors and enclosure for exit access stairways and exit access ramps	Rooms and enclosed spaces[c]
A-1 & A-2	B	B	C	A	A[d]	B[e]
A-3[f], A-4, A-5	B	B	C	A	A[d]	C
B, E, M, R-1, R-4	B	B	C	A	B	C
F	C	C	C	B	C	C
H	B	B	C[g]	A	A	B
I-1	B	C	C	A	B	B
I-2	B	B	B[h,i]	A	A	B
I-3	A	A[j]	C	A	A	B
I-4	B	B	B[h,i]	A	A	B
R-2	C	C	C	B	B	C
R-3	C	C	C	C	C	C
S	C	C	C	B	B	C
U	No Restrictions			No Restrictions		

Table 9-1. Interior wall and ceiling finish requirements by occupancy[k].

For SI:1 inch = 25.4 mm, 1 square foot = 0.0929 m².

[a] Class C interior finish materials shall be permitted for wainscotting or paneling of not more than 1,000 square feet of applied surface area in the grade lobby where applied directly to a noncombustible base or over furring strips applied to a noncombustible base and fireblocked as required by Section 803.4.1.

[b] In exit enclosures of buildings less than three stories in height of other than Group I-3, Class B interior finish for unsprinklered buildings and Class C interior finish for sprinklered buildings shall be permitted.

[c] Requirements for rooms and enclosed spaces shall be based upon spaces enclosed by partitions. Where a fire-resistance rating is required for structural elements, the enclosing partitions shall extend from the floor to the ceiling. Partitions that do not comply with this shall be considered enclosing spaces and the rooms or spaces on both sides shall be considered one. In determining the applicable requirements for rooms and enclosed spaces, the specific occupancy thereof shall be the governing factor regardless of the group classification of the building or structure.

[d] Lobby areas in Group A-1, A-2 and A-3 occupancies shall not be less than Class B materials.

[e] Class C interior finish materials shall be permitted in places of assembly with an occupant load of 300 persons or less.

[f] For places of religious worship, wood used for ornamental purposes, trusses, paneling or chancel furnishing shall be permitted.

[g] Class B material is required where the building exceeds two stories.

[h] Class C interior finish materials shall be permitted in administrative spaces.

[i] Class C interior finish materials shall be permitted in rooms with a capacity of four persons or less.

[j] Class B materials shall be permitted as wainscotting extending not more than 48 inches above the finished floor in corridors.

[k] Finish materials as provided for in other sections of this code.

[l] Applies when the exit enclosures, exit passageways, corridors, or rooms, and spaces are protected by a sprinkler system installed in accordance with Section 903.3.1.1 or 903.3.1.2.

Source: *2012 International Building Code.* Copyright 2009, Washington, DC: International Code Council, Inc. Reproduced with permission. All rights reserved.

Decorative materials that may change with the seasons are closely regulated by the model fire codes. The potential fire problem associated with decorative materials and holidays cannot be overstated. Christmas trees were the first item ignited in an estimated 360 structure fires per year, resulting in 15 civilian deaths, 44 civilian injuries, and over 16 million dollars in direct fire loss between 1999 and 2002.[6]

The model fire codes limit combustible decorative materials based on use, occupancy, and sprinkler protection. Inspectors must be particularly vigilant because many building owners don't see the potential fire hazard and may even be unaware that combustible decorative materials are regulated. Building owners are required to provide evidence that materials are noncombustible or flame-resistant (see figure 9-5).

NFPA 705, *Recommended Practice for a Field Flame Test for Textiles and Films* referenced by the legacy model fire codes is not referenced by the IFC and UFC. The field test was formerly contained in NFPA 701, *Standard Methods of Fire Tests for Flame Propagation of Textiles and Films* but was removed and became a stand-alone document in 1997. The NFPA 701 committee determined it was appropriate to create the field test as a stand-alone document, since the field test does not incorporate more rigorous laboratory testing methods incorporated into NFPA 701 testing such as conditioning of specimens, reproducibility, and repeatability. NFPA 705 continues to be maintained, with a prominent caveat the there is no correlation between the test and the testing prescribed by NFPA 701. The code and standards writers have left the inspector without a tool to gauge the relative safety of decorative materials in the field. Use the field test at your own peril or inform the business owner or organization to provide non-existent flame spread certification.

Flame resistant does not mean that a material will not burn. It means that if exposed to a small ignition source such as a cigarette lighter for a few seconds, the material burns little and self-extinguishes after the ignition source is removed. Treatments that render materials flame resistant must be renewed as needed.

UNSAFE STRUCTURES AND PROPERTIES

Vacant Structures

Vacant, dilapidated structures (see Figure 9-6) have a significant effect on the community. They become centers for illegal activity and vandalism and negatively impact the quality of the surrounding area. Property values drop, crime increases, people move out, and more properties are abandoned. Some would argue that the best thing that can happen in such instances is for the dilapidated buildings to burn down before the rest of the community is affected.

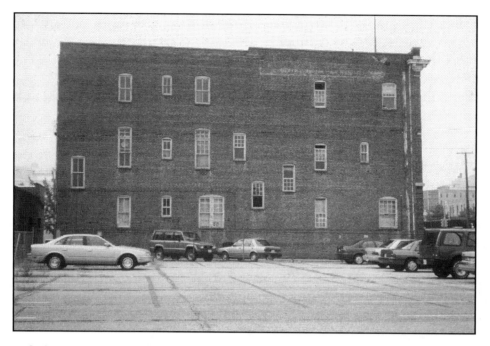

Figure 9-6. Vacant, dilapidated structures have a significant effect on the community.

The model fire codes term such buildings unsafe structures and require fire officials to refer such buildings to the building official for abatement, repair, or demolition. In many communities, building officials are limited to requiring owners to simply board up the first floor of such structures. Fire inspectors should be particularly vigilant in referring such structures and in following up to see that action is taken. We have a vested personal interest. It is the fire department that will be called on to extinguish fires in these structures. In December 1999, six Worcester, Massachusetts, fire fighters were killed while operating at a vacant cold storage warehouse. Eleven minutes into the fire, a neighbor advised the incident commander that a homeless couple had been living in the building. Twenty-two minutes later, one of the rescue crews became disoriented on the fifth floor and began to run low on air as fire conditions deteriorated. The crew was killed attempting to rescue people who were not in the structure at the time.[7]

Outdoor Waste Accumulation

The accumulation of wastepaper, wood, hay, weeds, or any combustible waste or rubbish in quantities that constitute a fire hazard (see Figure 9-7) are prohibited by the model fire codes. In order to enforce these provisions, the fire inspector must remember that the intent of the model code groups in developing the fire codes is to prevent the spread of fire; not to ensure that premises are neat in appearance, rodent free, or any of the other legitimate things that governments attempt to regulate.

An overgrown wood scrap pile in a backyard might be an eyesore for the neighbors. It might harbor rodents. It might be the very item that has caused a tiny war within a neighborhood. But it might not be a violation. All that one of the participants needs is some reinforcements (like you), to "win" the battle. Many times the condition ceases to be the real issue. The battle becomes the issue.

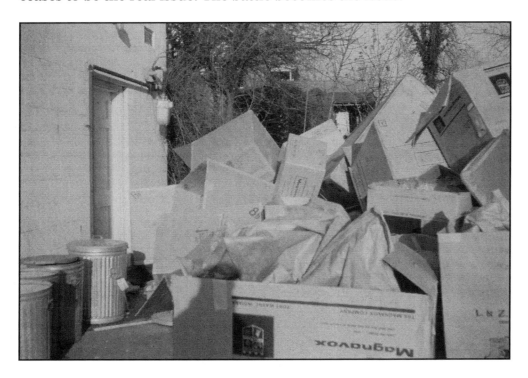

Figure 9-7. "The trash service only comes once a week," is not an acceptable excuse for this unsafe condition.

Such instances require a cool, detached assessment of the situation, then decisive action on the part of the inspector. Do not allow yourself to become involved in a neighborhood dispute. Take the necessary action to gain compliance with the code. Violations of the code must be addressed. Remember, you cannot make everybody happy. You can, however, gain the public's respect by fairly enforcing the fire code, referring matters outside your jurisdiction to the proper officials, and keeping all parties informed.

NOTE

You cannot make everybody happy; you can, however gain the public's respect by fairly enforcing the fire code.

The relative hazard of overgrown vegetation varies from relatively minor in most of the eastern United States, where rain is plentiful to severe in many parts of the country, not just the West and Southwest. The *International Fire Code* references the *International Urban-Wildland Interface Code* for vegetation clearance requirements in urban-wildland interface areas. The *Uniform Fire Code* includes similar requirements and references NFPA 1144 *Standard for the Protection of Life and Property from Wildfire*.

Electrical Hazards

The installation of electrical services, wiring, fixtures, and permanent equipment is regulated by the electrical code official of the jurisdiction. NFPA 70, the *National Electrical Code®* (NEC®), is adopted throughout most of the United States. The ICC's *International Electrical Code* establishes administrative text necessary to administer and enforce NFPA's *National Electrical Code®*, or NEC®, and complies with electrical provisions contained in the other international codes.

Electrical safety is achieved through four general objectives. The fire codes: (1) require that permanent electrical systems be maintained as originally approved; (2) limit the use of temporary wiring; (3) prohibit the use of hazardous appliances, equipment, and wiring (see Figure 9-8); and (4) ensure access to electrical service equipment

Figure 9-8. This open junction box/temporary wiring in an exit passageway of an occupied mall poses a hazard to the occupants.

Temporary Wiring

The use of temporary wiring is limited to construction, renovation, and demolition projects, and for special events and holiday periods not to exceed 90 days. Outdoor Christmas decorative lighting, carnivals, and similar activities are legitimate uses of temporary lighting. Provisions for temporary wiring are included in the NEC®. In most jurisdictions, an electrical permit must be secured prior to the installation of temporary wiring, and the installation must be inspected. The fire inspector's job is to see that it was installed under permit and is maintained.

Extension cords, power strips, and multiplug adapters are encountered by inspectors on a daily basis. The fire codes and the NEC® specifically address their use: Extension cords and power strips are only to be used to provide power to portable appliances. Freezers, refrigerators, and the like are not portable appliances.

- Only grounded extension cords may be used for grounded appliances.
- The ampacity of the extension cord must equal or exceed the rated capacity of the appliance.
- Extension cords may not be affixed to the wall or ceiling, nor passed through walls, ceilings, or doorways, nor under rugs or carpeting.
- Except for approved multiplug type cords, only one appliance may be powered by an extension cord.

Multiplug and cube-type plug-in adapters that do not comply with the NEC are specifically prohibited. Cube adapters, which enable building occupants to plug four or more appliances into a single receptacle, can easily lead to an overheating situation.

Hazardous Equipment

Appliances and equipment that are defective or have become unsafe due to lack of maintenance, misuse, or other damage must be removed from service. Fire inspectors should issue written orders and affix a tag prohibiting the use of the appliance or equipment. The electrical code official should be immediately notified.

Access to Electrical Service Equipment

Access to electrical service equipment is required for all occupancies. Clearance of at least 30 inches is required. Illumination is also required for the equipment service areas.

SMOKING

Smoking is a national health issue, with significant political implications. The fire codes only address the fire safety aspect of smoking. Smoking and the improper disposal of smoking materials have played a significant role in our national fire experience for many years. The most common fatal fire cause in residential occupancies continues to be improper handling/disposal of smoking materials.

In addition to the prohibitions against smoking contained within the fire codes, many state and local governments prohibit smoking in business and mercantile occupancies, restaurants, and other public buildings. Most of these ordinances are intended to protect the public from the nuisance and health hazards associated with secondhand smoke.

Citizen complaints regarding these ordinances are often directed to the fire prevention bureau. Informing an irate citizen that it is "not our problem," but you are really not sure who enforces the regulations is really bad. Trying to enforce regulations for which you have no legal authorization is worse. Citizen complaints should quickly be referred to the appropriate enforcing agency.

The fire codes specifically prohibit smoking in areas where flammable liquids, explosives, and other similarly hazardous materials are stored, used, or handled. These provisions are found within the specific chapters or articles of the code. The fire official is also empowered to order the posting of no smoking signs and the designation of specific safe smoking areas where conditions warrant. Piers, wharves, and storage, and industrial facilities all are likely candidates.

Smoking can also be prohibited during dry seasons in areas where there is significant potential for brush and wildland fires. Both the International Fire Code and the Uniform Fire authorize the fire official to prohibit smoking in outdoor areas where hazardous environmental conditions exist.

NOTE

Both the International Fire Code and the Uniform Fire authorize the fire official to prohibit smoking in outdoor areas where hazardous environmental conditions exist.

OPEN FLAMES

Open flame devices such as heating and lighting appliances, and torches for soldering pipe joints or removing paint are regulated within the general fire safety provisions of the fire codes. Cutting and welding using acetylene is addressed within a specific chapter or article in each code.

The use of open flame devices in areas where readily combustible materials are stored, used, or handled is prohibited. Where torches are used for soldering or removing paint, a capable fire watcher with an approved extinguisher must be posted and remain on the scene for at least 30 minutes after the operation has been completed.

NOTE

Where torches are used for soldering or removing paint, a capable fire watcher with an approved extinguisher must be posted.

Heating appliances, vents, and chimneys must be installed and maintained in accordance with the building and mechanical codes and with their listings. The fire codes require the sealing of unsafe appliances by the fire official. The use of portable fuel fired heating appliances is regulated by use group, depending upon the model code.

FIRE DEPARTMENT ACCESS

Emergency access on all-weather driving surfaces is required by the IFC and UFC for all commercial buildings and most residential subdivisions. Designated fire lanes are required to provide access to within 150 feet of all portions of the exterior walls of the first story, measured along the route of vehicular travel (see Figure 9-8). Requirements for width, vertical clearance, maximum grade, and turning radius are also specified. Private bridges are required to meet the American Association of State Highway Transportation Officials (AASHTO) *Standard Specification for Highway Bridges* standard.

Figure 9-8. Clear, posted fire lanes are critical to successful fire suppression and rescue operations. (Courtesy of Carl Maurice)

NOTE

Designated fire lanes are required to provide access to within 150 feet of all portions of the exterior walls of the first story, measured along the route of vehicular travel.

Inspectors should be on the lookout for security gate installations that have been installed without the necessary permits and approvals. The realization that fire apparatus cannot negotiate a turn into a subdivision because security gates have been installed isn't a good thing when there is a fire. The codes require adequate clearance as well as an approved means for emergency operation.

Requirements for fire apparatus access roads were slightly different under some of the legacy fire codes -- all the more reason to maintain copies of every code that has been adopted by your jurisdiction for future reference.

Summary

The model fire codes prescribe general regulations that apply to all occupancies. These regulations are aimed at preventing ignition and limiting the impact of fire in the event of ignition. To prevent ignition, smoking and the use of open flames are regulated. Electrical equipment must be maintained and used in accordance with the listing. The impact of fires can be reduced through the maintenance of rated construction elements, by limiting combustible decorative material and combustible waste, and by providing adequate emergency access for the fire department.

Review Questions

1. List the two basic strategies used by the model fire codes to safeguard life and property.

    a. _____
    b. _____

2. List three methods of preventing the ignition of fires by segregating the elements of the fire triangle.

    a. _____
    b. _____
    c. _____

3. List five methods of managing the impact of fire.

    a. _____

b. _____
c. _____
d. _____
e. _____

4. Which violation of the fire code presents the largest potential for the number of safety deficiencies that can exist?

_____

5. When dealing with an illegal change in use, what three actions should be taken by the fire inspector?

a. _____
b. _____
c. _____

6. List two barriers to fire that are not rated assemblies.

a. _____
b. _____

7. List two performance features of side-swinging fire doors.

a. _____
b. _____

8. Under what circumstances are hold-open devices permitted on side-swinging fire doors? _____

9. What code is referenced by the IFC to regulate the clearance of vegetation in urban-wildland interface areas? _____

10. List four general rules regarding the use of extension cords.

a. _____
b. _____
c. _____
d. _____

Discussion Question

1. During an inspection of a three-story office building, you NOTE that mail slots have been installed in all the doors to the tenant suites. The doors open into the main exit stairway for the building and all are listed fire doors.

    a. Has the code been violated?
    b. What information would you request from the building owner?
    c. What is the potential significance of this situation?

KEY TERMS

approved

acceptable to the code official with jurisdiction; approval is normally based on nationally recognized standards or, in their absence, on sound engineering practice

Chapter 10

Maintenance of Fire Protection Systems

Learning Objectives

Upon completion of this chapter, you should be able to:

- List the appropriate standards for the maintenance and inspection of sprinklers, standpipes, wet and dry chemical extinguishing systems, and fire alarm systems.
- Describe the importance of examining maintenance and inspection records kept by building owners.
- Describe what circumstances would cause an inspector to require an evaluation of an engineered fire protection system.
- Given flow test data, graph the available fire flow for a site.

INTRODUCTION

Fire protection systems play an important role in efforts to protect the public through fire safety and building regulations. The model codes recognize the potential effectiveness of these systems, which suppress, extinguish, or give early notification in fire situations. The codes reward building owners who install them by permitting increases in building height and area and even reduce fire-resistance ratings based on the supposition that the fire protection systems will function as designed and installed, for the life of the building.

The model building codes are designed to protect us from our own actions as well as those of our neighbors. The fire wall between town houses protects each neighbor from the other. If I am crazy enough to work on my motorcycle in the living room, the fire wall between us is the best friend you, as my next-door neighbor, have. If the rating on the wall has been reduced because the developer installed residential sprinklers, the scenario changes a bit. You now need the fire separation wall and my sprinkler system to protect you from my unsafe actions. Expecting a guy who does major motorcycle repairs in his living room to maintain a sprinkler system might be taking a lot for granted.

> **NOTE**
>
> **The model building codes are designed to protect us all from our own actions, as well as those of our neighbors.**

With fire protection systems, all of our eggs are in the maintenance basket. Out of service or impaired systems can also give a false sense of security to the occupants, causing them to needlessly endanger themselves by attempting to extinguish a fire, or by not promptly evacuating. The effective maintenance of fire protection systems is a primary goal of every fire prevention inspections program.

> **NOTE**
>
> **The effective maintenance of fire protection systems is a primary goal of every fire prevention inspections program.**

INSTALLATION AND MAINTENANCE STANDARDS

All fire protection systems and equipment are installed under a standard prescribed by the building or fire prevention codes and are required by the fire prevention codes to be maintained. With the exception of water-based extinguishing systems, maintenance provisions for fire protection systems are included in the standards for installation, many of which are found in Table 10-1.

Table 10-1. Partial list of fire protection system standards.

NFPA 13	*Standard for the Installation of Sprinkler Systems*
NFPA 13D	*Installation of Sprinkler Systems in One- and Two-Family Dwellings and Manufactured Homes*
NFPA 13R	*Installation of Sprinkler Systems in Low Rise Residential Occupancies*
NFPA 12	*Standard on Carbon Dioxide Extinguishing Systems*
NFPA 12A	*Standard on Halon 1301 Extinguishing Systems*
NFPA 17	*Standard for Dry Chemical Extinguishing Systems*
NFPA 17A	*Standard for Wet Chemical Extinguishing Systems*
NFPA 20	*Standard for the Installation of Stationary Pumps for Fire Protection*
NFPA 24	*Standard for the Installation of Private Fire Service Mains and their Appurtenances*
NFPA 72	*National Fire Alarm and Signaling Code*
NFPA 14	*Standard for the Installation for Standpipe and Hose Systems*
NFPA 10	*Standard for Portable Fire Extinguishers*
NFPA 25	*Standard for the Inspection, Testing, and Maintenance of Water-Based Fire Protection Systems*

NFPA 17, *Standard for Dry Chemical Extinguishing Systems* includes the installation and maintenance requirements for dry chemical systems. In 1995, NFPA 25, *Standard for the Inspection, Testing and Maintenance of Water-Based Fire Protection Systems* was issued. The standard incorporated requirements for all water-based extinguishing systems including private fire mains, fire pumps, and storage tanks. Detailed maintenance and inspection procedures as well as sample forms and documents are included in the standard.

Figure 10-1. This dry chemical extinguishing system must be maintained in accordance with NFPA 17 *Standard for Dry Chemical Extinguishing Systems*. (Courtesy of A. Maurice Jones)

WATER-BASED EXTINGUISHING SYSTEMS

Records of all fire protection system inspections, tests, maintenance, and repairs must be maintained on the premises by the owner or occupant. All test results, defects, and corrective actions must be logged. An important part of every routine inspection is a review of all fire protection system records for the building. Records should be well organized and up-to-date. They reflect the operational readiness of the fire protection systems in the building. The lives of the building occupants, as well as the firefighters who respond to an incident in the building, may depend on these systems.

> **NOTE**
>
> **An important part of every routine inspection is a review of all fire protection system records for the building.**

NFPA 25 (2014) Appendix B describes five formats for test records. It directs users to develop their own forms based on their particular systems or to purchase forms from the American Fire Sprinkler Association or National Fire Sprinkler Association. Forms provided by the two fire-sprinkler trade associations are checklists that take the guessing out of what needs to be done and how often.

Sprinkler Systems

Sprinkler systems are very reliable. In a 1985 Canadian Building Digest article, J. K. Richardson placed the reliability of sprinkler systems at greater than 96 percent, based on statistics compiled by the National Fire Protection Association, the Australian Fire Protection Association, and the city of New York. The statistics were maintained from the late 1800s through the 1960s.[1] NFPA discontinued tracking the performance of sprinklers in 1970, citing that the data collection was biased toward poor performance, which could result in an inaccurate perception of sprinkler effectiveness.[2] Nonetheless, information from statistics collected prior to 1970 clearly identifies areas that require heightened vigilance by owners, design professionals, contractors, and code officials. Table 10-2, lists the top six reasons cited for unsatisfactory from 1925 to 1969.[3]

Table 10-2. Leading causes for unsatisfactory sprinkler performance 1925–1969.

Problem	Percentage of cases
Water shut-off	35.4%
Inadequate water supply	9.9%
System inadequate for level of hazard	13.5%
Inadequate maintenance	8.4%
Obstruction to water system	8.2%
System only designed for partial protection	8.1%

Source: Kimberly D. Rohr, "*U.S. Experience with Sprinklers,*" NFPA 2001, p. 48.

It doesn't take a rocket scientist to see that the primary causes for unsatisfactory performance prior to 1970 were issues that could be readily identified by inspection and testing. In 1970, most people in fire protection never envisioned the sprinkler recalls that began to occur in the 1990s. The beauty of sprinklers was their simplicity, some of which faded with the development of high-performance heads such as Central Sprinkler's Omega.

The inspection of every building equipped with a sprinkler system should include an inspection of the sprinkler riser and all control valves (see Figure 10-2). Closed valves are the primary cause of failure of water-based extinguishing systems in protected occupancies. Begin at the riser where the water supply riser enters the building. All sectional control valves should be visually inspected to ensure that they are open.

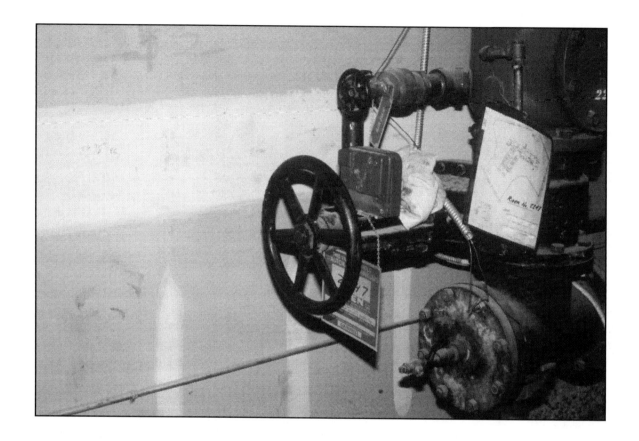

Figure 10-2. Closed valves are the primary cause of failure of water-based extinguishing systems in protected occupancies.

A hydraulic nameplate that is required to be attached to the riser on all hydraulically calculated systems. The nameplate must indicate the design area, discharge density (see Figure 10-3), required flow and residual pressure demand at the base of the riser, and the hose stream demand. As we discussed in Chapter 6, discharge density, or the amount of water discharged over a given area, is based on the occupancy hazard

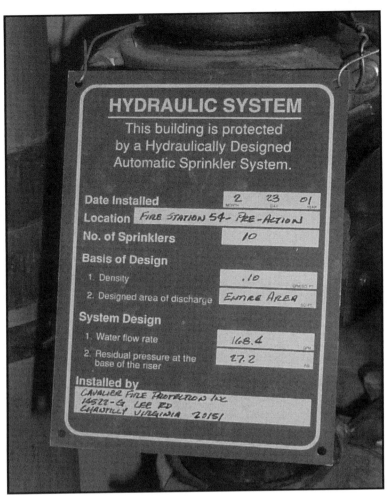

Figure 10-3. A durable hydraulic nameplate is required to be permanently affixed to the riser of every hydraulically calculated sprinkler system. (Courtesy of A. Maurice Jones)

classification established in NFPA 13, Standard for the Installation of Sprinkler Systems (see Table 10-3). The inspector should determine that the processes within the building are consistent with the occupancy hazard classification under which the sprinkler system was designed.

NOTE

The inspector should determine that the processes within the building are consistent with the occupancy hazard classification under which the sprinkler system was designed.

Table 10-3. Occupancy classification examples from NFPA 13. (Source NFPA 13, 2014)

CLASSIFICATION	EXAMPLES
Light Hazard	Churches, hospitals, schools, offices, theaters, residences
Ordinary Hazard Group 1	Parking garages, canneries, laundries, restaurant service areas, electronic plants
Ordinary Hazard Group 2	Dry cleaners, horse stables, machine shops, post offices, print shops, library stack rooms
Extra Hazard Group 1	Aircraft hangars, sawmills, plywood manufacturing
Extra Hazard Group 2	Flammable liquid spraying, plastic processing, varnish and paint dipping

If the hydraulic nameplate indicates that a system has been designed to discharge at a density of 0.15 gpm/ft.^{2} over a 1,500 square foot area, the system has been designed at the Ordinary Group 1 curve. If the building is being used as a printing facility, the occupancy hazard classification from Table 10-3 has obviously changed, and the system must be reevaluated for the greater hazard. It must now be recalculated at the Ordinary Group 2 curve. A notice of violation should be issued requiring an evaluation of the system, and, if indicated, requiring an upgrade of the sprinkler system to protect the current hazard.

While in the sprinkler room, the stock of spare heads should be inspected and the heating appliances should be checked. Dry pipe valves must be maintained at a minimum 40°F. Wet pipe systems must also be maintained at a minimum 40°F. Frozen pipes (see Figure 10-4) unnecessarily endanger building occupants by voiding the sprinkler protection and do considerable water damage. Water damage from frozen pipes can cause entire communities to develop strong feelings against sprinklers. The inspection of insulation, heating appliances, and methods of preventing trapped water in dry piping are every bit as important as checking the supply valves.

Figure 10-4. This frozen pipe was removed from the unheated attic of an assisted living facility where tradesmen disturbed insulation, exposing the pipe to freezing temperatures.

NOTE

Water damage from frozen pipes can cause entire communities to develop strong feelings against sprinklers.

NFPA 25 includes provisions for visual inspection of components, the test and manipulation of valves and controls, the removal and test of antifreeze solutions, and the removal and laboratory test of heads according to a specific schedule. Corporations with effective safety and risk management programs and those whose insurance companies retain firms such as FM Global (Factory Mutual Engineering) often meet or exceed the requirements for maintenance and testing. Most others only recognize the importance of sprinkler testing and maintenance with the assistance of the fire official.

Standpipe Systems

Visual inspection and testing are required by NFPA 25 for standpipe and hose system components. Five-year flow tests in which water is flowed from the highest or most remote point in the system should never be waived due to inconvenience. Flow tests should not be conducted during freezing weather due to the obvious safety concerns, but should be conducted when temperatures moderate. Flow tests are the most effective means of determining a continued adequate water supply and of effectively testing the entire supply system from the street to the topmost outlet. An effective inspection system will identify water supply problems caused by obstructions and such problems as partially closed valves in the municipal water system. Closed valves, foreign material, and even water mains that have been crushed by heavy equipment may lurk beneath the asphalt and concrete. Water supply problems are not uncommon occurrences. They must be detected and promptly resolved.

NOTE

Flow tests are the most effective means of determining a continued adequate water supply and of effectively testing the entire supply system from the street to the topmost outlet.

Other components that require testing and inspection include pressure regulating devices, which must be tested every 5 years to ensure that adequate flow and pressure is provided at each hose outlet. Standpipe hose and nozzles must also be tested.

Fire Pumps

Fire pumps might be considered the most critical element in many water-based extinguishing systems. While all portions of the system are important, a lot more can go wrong with a fire pump than with piping, valves, and tanks. The lack of maintenance or misuse can lead to expensive repairs that some owners would just as soon forego at the risk of disabling the standpipe and sprinkler systems. Fire pumps must be run and inspected weekly and subjected to an annual flow test. Your inspection of the fire pump includes an inspection of the building engineer's records and his knowledge of the fire pump. Time spent educating building engineers and assisting them in complying with the code is time well spent.

Other water-based systems and components such as fixed water spray and foam as well as water tanks, private fire service mains, and fire department connections (see Figure 10-5) have fixed maintenance and inspection schedules specified by NFPA 25.

Figure 10-5. Ongoing construction has made this fire department connection inaccessible, exposing firefighters to unnecessary hazards. A temporary FDC should be located at an accessible location.

FIRE FLOW TESTS

Flow testing of hydrants is the most efficient and effective method of verifying an adequate water supply for fixed fire suppression systems and of ensuring adequate water supply for manual fire extinguishment by the fire department. Unfortunately, it is a function that has, in large part, gone by the wayside due to the constantly expanding list of responsibilities that many modern fire departments have undertaken (See Figure 10-6). Whether the fire department actually conducts flow tests or merely reviews the test data provided by another organization, knowledge of proper test practices and procedures and the ability to "understand the picture" when test data is presented in graphic form are fundamental skills for the fire inspector.

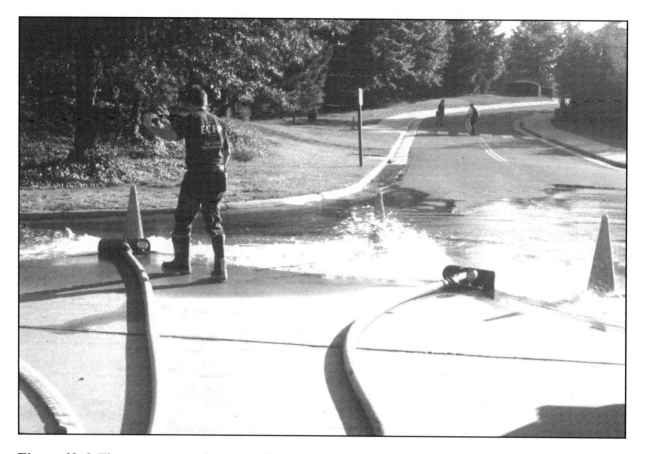

Figure 10-6. Fire pumps must be run and inspected weekly and subjected to an annual flow test.

> **NOTE**
>
> **Whether the fire department actually conducts flow tests or merely reviews the test data provided by another organization, knowledge of proper test practices and procedures and the ability to "understand the picture" when test data is presented in graphic form are fundamental skills for the fire inspector.**

The term flow testing includes routine tests from main drains and inspector's test pipes, flows from the topmost or most remote standpipe outlets, fire pump flow tests, and hydrant and water supply system tests. There are three major points to keep in mind about flow tests:

1. The frequencies for testing contained within the model codes and referenced standards are minimum frequencies not optimal frequencies. If they are not being performed at least at the required frequency, trouble could be brewing.

2. Testing accomplishes more than just determining that the same amount of water flows through the discharge orifice from one test to the next. Tests provide sorely needed flushing of sediment and foreign material from mains and piping.
3. Testing of municipal water systems is the best way for firefighters to learn about the municipal water system, and where the best flows and the worst flows are available. This knowledge can be of vital importance. In the early stages of major incidents, the road system around major incidents quickly becomes blocked by vehicles, making efficient hose lays almost impossible after initial companies have been positioned. Adequate water supply may depend on a connection to the municipal water system at a point with adequate main size, not necessarily the main that is adjacent to the fire incident.

The NFPA 1031 (Professional Qualifications for Fire Inspector) committee recognized the trend away from hands-on flow testing in the fire service, which is reflected in the Job Performance Requirement (JPR) for Fire Inspector I (2014):

Verify fire flows for a site, given fire flow test results and water supply data, so that required fire flows are in accordance with applicable codes and standards and all deficiencies are identified, documented, and reported in accordance with the policies of the jurisdiction.

Whether your department actually performs flow tests or not you need to be familiar with proper test procedures to be able to evaluate test data submitted by other organizations. NFPA 291, *Recommended Practice for Fire Flow Testing and Marking of Hydrants* should be consulted for guidance.

Flow Test Considerations

Flow tests must be conducted safely. Care should be taken to prevent property damage and to minimize the impact on adjacent businesses and residences. Flow tests must be coordinated with the local water authority. Property damage ranging from the interruption of a manufacturing process to ruining a family's wash with rust and scale can be avoided with a little planning.

Always perform a safety check prior to conducting a flow test. The safety of the public and of fire department personnel must be your first priority. Consider these items:

- Weather conditions: Will the water drain to a safe location prior to freezing? Creating icing conditions on streets, sidewalks, or parking lots is simply not an acceptable result of a routine testing procedure, conducted in the name of safety.
- Time of day: Will the test disrupt traffic during rush hour or create a traffic hazard? Will the test negatively impact business operations or cause hardship to the occupants of adjacent residences? Could children walking to school be impacted?

- Topography: Where will the water drain? Are there open manholes or underground vaults, construction operations, or utility workers in the vicinity? Could runoff endanger workers or damage equipment?
- Location: Are hydrants situated so that flowing water strikes automobiles or pedestrians, damages landscaping, or undermines property?

The potential benefits from conducting flow tests are enormous. The potential for damage, injury, or a public relations nightmare is real. Proper planning and coordination will minimize the possibility of a problem.

Flow Test Equipment

In order to conduct a hydrant flow test, the following equipment is needed:

- A good quality (preferably steel) ruler with 1/16th inch divisions
- A pitot tube with a bourdon gauge, or a flow meter
- A 2 1/2-inch hydrant cap, with a bourdon pressure gauge and petcock blow-off
- A hydrant wrench(s)
- Writing paper and pencil to record test data and sketch the area

Flow Test Procedures

The test hydrant is selected due to its proximity to a particular building or process. The test procedure is designed to determine the available flow from the water system at the test hydrant. Flowing of the test hydrant is limited to flushing it to remove sediment and debris, and to ensure an adequate pressure reading. A flow hydrant is downstream from the water source and test hydrant and should be selected to minimize potential property damage and traffic disruption. Refer to a map of the water system or check with the water authority to determine direction of flow within the water system.

When operating fire hydrant valves, open and close them slowly to prevent water hammer. Always open them fully, then, if necessary, throttle down by closing several turns. *Note for proper drainage in dry hydrants.* Blocked drains will lead to frozen hydrants and must be immediately reported to ensure their timely repair. Always check for damage to threads, obstructions such as shrubbery or fences, and damage. Report these conditions to the proper authority.

At the test area, after making the appropriate notifications and safety check, the following sequence should be followed:

1. Fully open and flush the test hydrant until it flows clear. Failure to flush the test hydrant may introduce sediment or debris into the gauge. Attach the 2 1/2-inch cap with pressure gauge. Fully open the test hydrant, open the petcock to bleed

off air from the barrel, and close the petcock. NOTE and record the pressure reading. This is the **static pressure**, or pressure exerted within the water system at no flow. Some waterworks technicians will refer to this a "line pressure," or pressure on the water lines. (See Figure 10-7.)

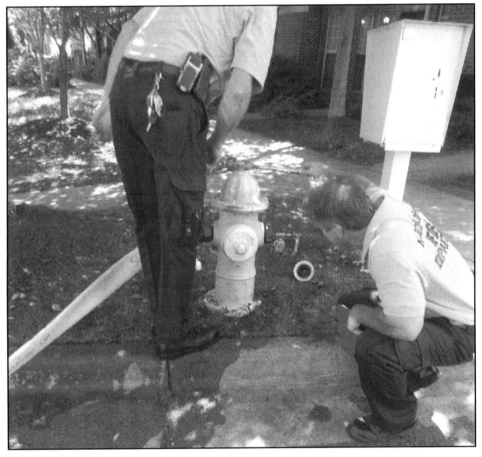

Figure 10-7. After flushing the test hydrant, open the valve and note the static pressure. (Photo courtesy of A. Maurice Jones.)

2. Remove a 2 1/2-inch cap from flow hydrant number one (additional flow hydrants may be required). Measure and record the inside diameter of the discharge Reach into the discharge and determine its shape. Then choose the hydrant **coefficient of discharge**, a number that describes the opening, as shown in Figure 10-8, and record it.

Figure 10-8. The coefficient of discharge is based on the style of hydrant and the reflects the friction loss due to the configuration of the discharge outlet.

3. Fully open the flow hydrant until it runs clear. Insert the pitot tube into the stream or note the pitot pressure as shown in Figure 10-9. Record the pressure indicated on the bourdon gauge; this is referred to as **pitot pressure**. It may take a few minutes for the pressure to stabilize. If readings fluctuate, use the average pressure.

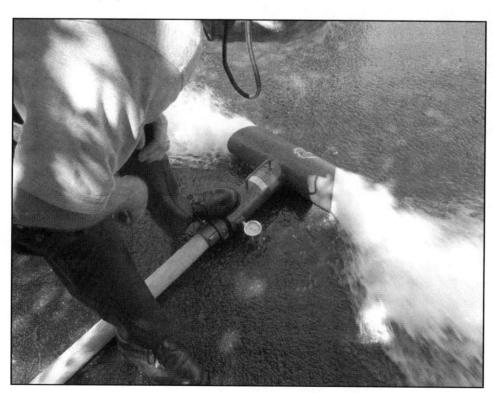

Figure 10-9. Record the flow pressure at the flow hydrant(s). (Courtesy of A. Maurice Jones)

4. While water is flowing from the flow hydrant(s) record the pressure indicated by the gauge on the test hydrant. This is the **residual pressure**. Additional downstream hydrants may have to be flowed simultaneously in order to achieve at least a 10-psi drop in pressure at the test hydrant.
5. Once satisfactory test information has been obtained, the flow hydrants can be slowly shut down and checked for proper draining. Any abnormality should be reported to the water authority.
6. When shutting down the test hydrant, the petcock on the 2 1/2-inch gauge should be opened. Failure to open the petcock allows water draining from the barrel to create a partial vacuum, potentially damaging the gauge. Ensure that the test hydrant drains.

Graphing Fire Flow Data

Information from flow tests can be plotted using logarithmic graph paper, entered into a computer program, or calculated using flow tables or formulas. To convert pitot pressure to flow, use the following formula:

Flow (in gpm) = $29.83 = D^2 \times C_d \times \sqrt{P}$

where D is the inside diameter of the discharge orifice, Cd is the coefficient of discharge for the hydrant type, and P is the pitot pressure.

Graphing flow data using logarithmic graph paper is a quick method to compare flows with those of previous years. All that is needed is flow test data, 1.85 exponential or logarithmic, graph paper, a straight-edge, and a very sharp pencil. (Mechanical pencils are extremely useful.)

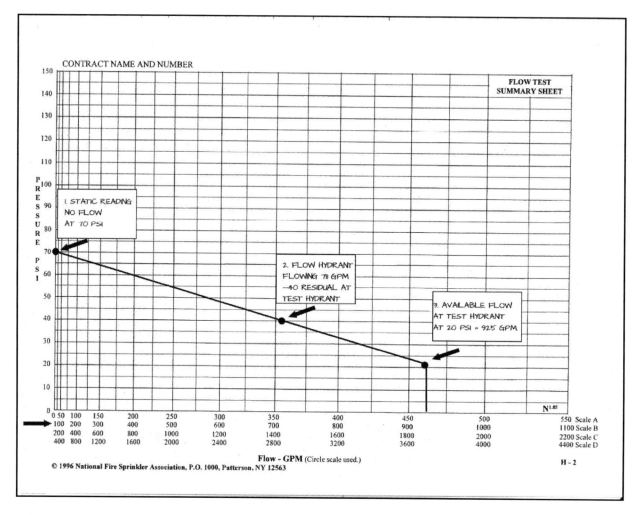

Logarithmic graph paper printed specifically for flow tests such as shown in **Figure 10-10** above is available from the National Fire Sprinkler Association and engineering supply companies. The x-axis (horizontal) represents flow; the y-axis represents pressure. NOTE that there are several scales for the x-axis to accommodate a wide range of flows. To compare test results with those of previous tests, it is important to u   se the same scale.

Figure 10-10 is an example of graphing fire flow data using the following steps:

1. Locate and mark the static pressure on the y-axis (vertical) at 0 gpm flow (point 1).
2. Locate the total test flow along the x-axis and the residual pressure along the y-axis. The point at which these values intersect (point 2) represents the residual pressure reading at the test hydrant at total flow from your flow hydrant(s). Mark this point.
3. Use a straightedge to connect the two points, extending the line from the y-axis base line to the x-axis base line (you can stop where it intersects 20 psi).

4. Locate the point at which the line you have just drawn intersects the 20-psi line along the y-axis (point 3.) This point on the x-axis is your total available flow at 20 psi.

The Value of Flow Testing

Fire flow from hydrants is by definition "the amount of water that is available at 20 psi residual pressure for firefighting." There is actually more water available, but operating fire department pumpers at lower residual pressures can result in a negative pressure at points within street mains. This negative pressure could cause the collapse of the mains or other water system components, or back-siphoning polluted water from some other interconnected sources. Operating at residual pressures of less than 20 psi (1.4 bar) is not permitted by many state health departments.

Current results should be compared with previous tests. Changes may reflect increased usage, the addition of equipment or capacity by the water authority, or obstructions such as foreign material or partially closed valves. Changes should be reported to the water authority for investigation and resolution. Valves that were closed for construction or maintenance and then not fully opened are a common but significant impairment that may only be identified by flow testing or failure of the system during a fire incident.

WET AND DRY CHEMICAL EXTINGUISHING SYSTEMS

Wet and dry chemical extinguishing systems are used to protect various industrial hazards from commercial cooking operations to spray-painting booths. The systems must be maintained in accordance with their respective standards (NFPA 17A and 17), the manufacturers' recommendations, and the fire prevention codes. The IFC and UFC require testing of all operating components every 6 months. Key items for the inspector to check are:

- Adequate system pressure
- Adequate coverage (appliances or process has not changed)
- Buildup of grease (see Figure 10-11) or residue on fusible links or elements
- Obstructions to mechanical operation of the system
- Obstructions to manual activation controls (see Figure 10-12)
- Placement of required supplementary fire extinguishers

Figure 10-11. This system is dripping grease, has unsupported piping and an unapproved mesh filter. The inspector should order a cessation of cooking operations until corrections are made.

In Chapter 6, we discussed the five classes of fire and some of the different agents used to extinguish them. Any mixing of extinguishing agents should be strictly avoided. Although some agents such as potassium bicarbonate dry chemical and aqueous film-forming foam have been found to be effective together, most are not. The different substances can react and reduce or even eliminate the effectiveness of one or both. A good example is the incompatibility between the two dry chemical agents potassium bicarbonate (BC) and ammonium phosphate (ABC).

NOTE

Any mixing of extinguishing agents should be strictly avoided.

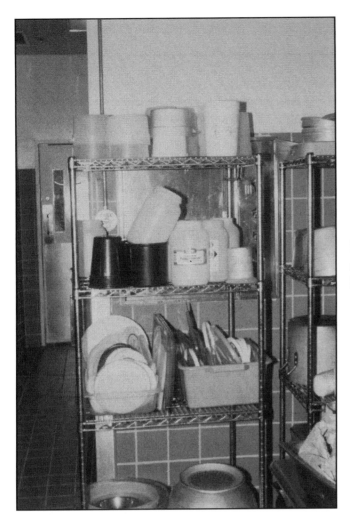

Figure 10-12. The manual actuator for the range hood suppression system is blocked by kitchen equipment.

Effective June 30, 1998, NFPA 10, Standard for Portable Fire Extinguishers has required the installation of a fire extinguisher with a Class K rating as supplemental protection for commercial cooking operations. The standard specifically exempts "extinguishers installed specifically for these hazards prior to June 30, 1998." (NFPA 10, 2-3.2) Any system installed or upgraded after June 30, 1998, must meet the new requirement. Fire equipment manufacturers are no longer manufacturing replacement parts for dry chemical, commercial cooking fire protection systems.

Inspectors must be on the alert for changes in the hazards that these engineered systems have been approved to protect. The addition of cooking appliances or even the upgrade of an appliance may tax a system beyond its design capabilities. There is very little fat in engineered fire protection systems. The replacement of a commercial fryer with a modern high efficiency unit that has multiple heating elements and 50 percent more cooking oil as a fuel load requires an engineering evaluation of the fire suppression system.

> **NOTE**
>
> **The addition of cooking appliances or even the upgrade of an appliance may tax a system beyond its design capabilities.**

Halon and Clean Agent Systems

NFPA Standards 12A and 2001, respectively, include installation and maintenance provisions for total flooding halon and clean agent fire extinguishing systems. System components (detection and initiating devices) must be inspected and tested at regular intervals. Tanks and containers must be tested every 5 years. This testing does not involve discharging the systems, only the operation of detection devices and actuator systems.

Fire Alarm Systems

In 1993, NFPA consolidated the various fire alarm and detection equipment standards into NFPA 72, *The National Fire Alarm Code*. The installation, inspection, testing, and maintenance of all fire alarm systems and equipment are included in NFPA 72. An effective program of enforcing the maintenance and inspection provisions for fire alarm systems will result in a decrease in responses to nuisance alarms by the fire department. Reducing the number of nuisance alarms not only saves time and money, but potentially saves lives as well. The lives of the public are better protected because companies spend less time out of service on nuisance alarms, and the firefighters and public face less exposure to the traffic hazards associated with emergency response.

> **NOTE**
>
> **Reducing the number of nuisance alarms not only saves time and money, but potentially saves lives as well.**

Summary

With few exceptions, the fire protection systems that provide building and life safety protection are engineered; that is, they are designed for a specific hazard. Standards for installation, maintenance, testing, and inspection are referenced in the model building and fire prevention codes. The job of the fire inspector is to ensure that the system is functional, that the maintenance and inspection provisions are being followed, and that the hazard the system was designed to protect has not changed.

Review Questions

1. What standard was developed for the maintenance of water-based extinguishing systems by the National Fire Protection Association?

2. List three items of information that should be included in fire protection system records.

   a. _____
   b. _____
   c. _____

3. What is the primary cause of failures in water-based extinguishing systems?

   _____

4. List four items that must be included on sprinkler system hydraulic nameplates.

   a. _____
   b. _____
   c. _____
   d. _____

5. List the five occupancy classifications from NFPA 13.

   a. _____
   b. _____
   c. _____
   d. _____
   e. _____

6. Wet pipe sprinkler systems must be maintained at a minimum of _____ degrees to protect against freezing.

7. What is the most effective method of determining an adequate water supply for an existing sprinkler system?

   _____

8. List six key items to be checked during an inspection of a wet or dry chemical extinguishing system.

   a. _____
   b. _____
   c. _____
   d. _____
   e. _____

f._____

9. Supplementary fire extinguishers installed in the vicinity of range hood fire suppression systems should be rated for Class _____ only.

10. What standard for the installation and maintenance of fire alarm systems was developed by the National Fire Protection Association? _____

Discussion Question

1. During an inspection of a restaurant, you notice that a new range and fryers have been installed. The old appliances had been in place since the establishment was first opened over 20 years ago. The restaurant owner informs you that the equipment salesman said they did not need a permit for installation because it was "replacement with like equipment."

   a. Does a code violation exist?
   b. What action should the fire inspector take in this case?
   c. Explain your actions to the chief who receives a call from a city councilman who claims you are harassing the restaurant owner.

KEY TERMS

static pressure

pressure exerted within a water system at no flow, also referred to as "line pressure."

coefficient of discharge

a number describing the orifice (opening) from which water flows from a hydrant

pitot pressure

the pressure (in psi) shown on the bourdon gauge attached to the pitot tube, inserted in a water stream flowing from a fire hydrant

residual pressure

the pressure (in psi) measured at the test hydrant with water flowing from the flow hydrant.

Chapter 11

Means of Egress Maintenance and Fire Safety and Evacuation Planning

Learning Objectives

Upon completion of this chapter, you should be able to:

- Name five ways that the means of egress from a building or space can be compromised.
- Describe the importance of posted occupant loads in assembly buildings.
- Describe the process used to establish the occupant load.
- Describe what constitutes "overcrowding" in an assembly occupancy.

INTRODUCTION

Throughout this text, we have discussed the model codes' balancing act with respect to building safety elements used in building design. Requirements for fireresistive construction features, fire protection, interior finish, and means of egress are based on the needs of the building occupants and on the relative hazards associated with the building use. In theory, at least, when buildings are designed and constructed in accordance with model codes, every element of safety and security is in perfect harmony. People are safe, happy, and productive.

Unfortunately, that's not always the case. From the time the certificate of occupancy is issued by the building official to the time the building is mercifully torn (or unmercifully burns) down, it is under a constant assault by the owners, occupants, and outsiders. In the name of doing things better, the balance created by the model codes is often lost. Unquestionably, of all the elements of building safety prescribed by the model building codes, means of egress is the easiest to compromise. It generally takes time and money to seriously affect other major building features.

It only takes an industrious store manager to realize that all that valuable space in the back hallway and stairwell is going to waste before the big Christmas sale. She dutifully fills the exit corridor and stairs with stocks of combustible goods. The overheated cook who props open the door leading into the exit corridor of the shopping mall is only attempting to get some ventilation. In cooling off the kitchen he puts hundreds of people at risk from a fire incident anywhere within the mall. Simple, everyday occurrences that compromise the means of egress combine with the not so common to threaten the occupants in the event of fire.

MAINTAINING THE MEANS OF EGRESS

The same elements involved in the original establishment of the means of egress must be revisited in order to assess the adequacy of the exits from any building or space. First remember the three parts of a means of egress (see Figure 11-1): exit access, exit, and exit discharge. Not all parts are regulated in the same fashion. The length of travel in the exit access is regulated, which is not the case in the exit or the exit discharge. Storage is not permitted within an exit, but is acceptable within most parts of the exit access.

> **NOTE**
>
> **The length of travel in the exit access is regulated, which is not the case in the exit or the exit discharge.**

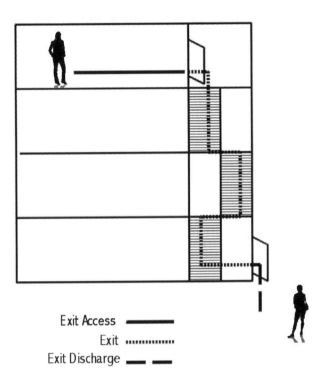

Figure 11-1. The 3 parts of the means of egress.

In Chapter 7, we saw that a six-step process can be used to design or assess the adequacy of the means of egress from a building or space. We used many of the same steps, in a different order, to determine the capacity of the different means of egress components in existing spaces. Fortunately, these exercises are not necessary on most inspections. Most of the things people do that compromise the means of egress fall into five broad categories. Exits are obstructed or obscured, the integrity of their construction is compromised, the integrity of their function is compromised, or the exit is eliminated. In order to perform an adequate inspection of any building or space every component of every means of egress must be checked by the inspector. The inspector must walk every stair and corridor and open every door. Without a methodical check of each component, the inspection is incomplete. The means of egress from a high-rise building with forty stories of perfectly operating fire doors is worthless if the stairway discharge door at the ground floor is blocked from the outside by shrubbery and cannot be opened.

NOTE

Every component of every means of egress must be checked by the inspector.

Obstructions

By far the most common violation regarding the means of egress is the blocking of doors, aisles, or passageways (see Figure 11-2). Often the condition is remedied by the occupants within minutes of occurrence. Other times the obstruction is permanent or semi-permanent. Restaurants commonly store high chairs or stacks of extra chairs in passageways that lead to rear exits, making them impassable. The same passageway, which probably also leads to the rest rooms, is a great place for the cigarette machine. Reducing the width of the passageway by the width of the cigarette machine probably creates a means of egress that falls below the code-required minimum width.

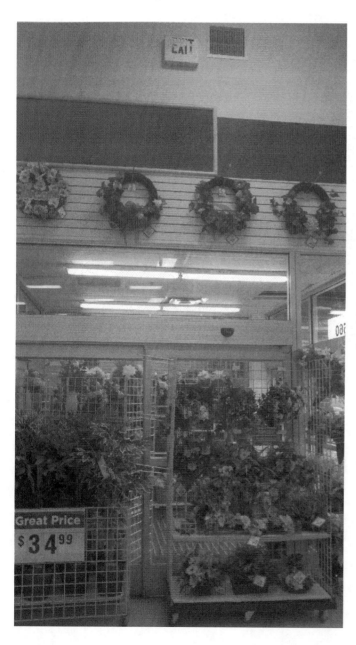

Figure 11-2. This pair of exit doors was blocked on the day of a big sale. (Courtesy of Jeremy Luttrell)

NOTE

By far the most common violation regarding the means of egress is the blocking of doors, aisles, or passageways.

Obvious obstructions such as storage are generally not difficult for the inspector to remedy. Most building owners and occupants can understand that doors and aisles should not be blocked. For them it is a matter of degree: They usually feel that they did not have it blocked that much and people could have gotten out if they really needed to.

Unfortunately, sometimes reason does not prevail. Some business owners insist on obstructing the means of egress, sometimes in order to accommodate additional stock and at other times for sheer convenience. Instances where large retail chains have repeatedly obstructed the means of egress during the holiday shopping season are numerous. Shopping malls and regional shopping centers should be frequently checked prior to and during the holiday shopping season with advice to management that all mercantile occupancies are being similarly inspected and to expect them to be checked again.

NOTE

Shopping malls and regional shopping centers should be frequently checked prior to and during the holiday shopping season.

Obstructions such as high chairs or other equipment in the rear corridor can be much more difficult. Small business owners often have few options due to limited space. "It has been like this for years and none of the other inspectors ever said anything," is also a common theme. If a vending machine or even fire protection equipment reduces the width of the passageway below the code-required minimum, you must order the owner to move it to another location.

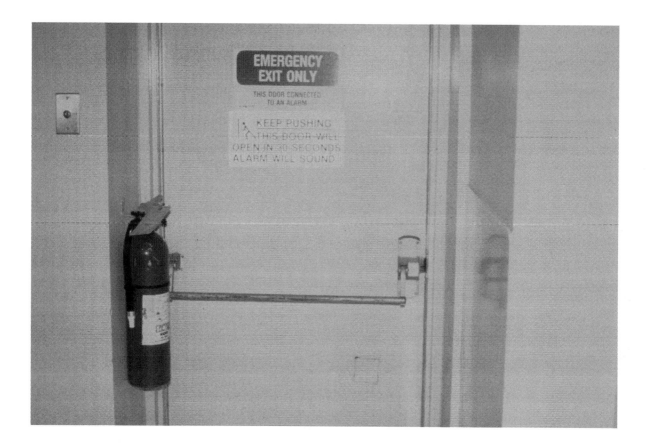

Figure 11-3. This fire extinguisher reduces the clear width of the exit and should be relocated.

Egress ways can also be obstructed without the knowledge or assistance of the owner or tenant. Snow and ice that have accumulated on exterior stairs, fire escapes, or in front of exterior doors can reduce or even eliminate entire exits. Trees or shrubs have been known to grow to such proportions that exterior doors are blocked, again without the knowledge of the building occupants.

> **NOTE**
>
> **Snow and ice that have accumulated on exterior stairs, fire escapes, or in front of exterior doors can reduce or even eliminate entire exits.**

Obscuration

Means of egress components are generally obscured by one of two methods. In the first, for whatever reason, the exit is hidden from view. Draperies used for privacy can camouflage doors, making them unavailable for the public. Large bookcases or display cases can also be arranged in ways that may not obstruct access to the exit, but which block any view of the exit making it just as unusable for the public.

The second method of obscuration does not involve disguising or hiding the means of egress as much as not showing the building occupants where the exits are. Illumination and marking of the means of egress with lighted exit signs is necessary if the public is to be expected to evacuate. With the exception of dwelling units and occupancies with small occupant loads and a single exit, most buildings and spaces will be equipped with internally illuminated exit lights and means of egress lighting connected to an emergency power source. The emergency power source may be a battery system or generator, but must pick up the assigned load within 10 seconds of power failure.

Emergency lighting is designed to enable the occupants to safely exit the building in the case of an emergency. The required intensity of 1 footcandle is not sufficient for much more than that. The 1–1 ½ hour duration required by the building codes is not designed to give the occupants time to finish shopping or eating before they evacuate the building.

Compromises of the Integrity of Construction Elements

Many elements of the means of egress have specific fire-resistance rating requirements or operational requirements. Exit access corridors that serve thirty or more people, exits, smokeproof enclosures, and horizontal exits all have specific requirements for the construction of structural assemblies and opening protectives. The integrity of these elements can be compromised by tradesmen who penetrate rated assemblies while performing utility work, or by a tenant who innocently props open the stairway door to make it easier to reenter with his hands full. Both scenarios are common occurrences and both can have the same effect of making some part of the means of egress less effective or perhaps ineffective.

If a fire occurs on the same floor where our friend has propped open the stairway door and there are only two stairwells, we could conceivably lose 50 percent of the exit capacity of our building. What if the building is a high-rise? These are all "what-ifs," but if what-ifs did not occur, there would be no need for building or fire codes.

Operational requirements include those for door hardware such as locking and latching, direction of door swing, and marking of interior stairway doors. Special locking arrangements and access-controlled egress doors pose special problems in large office buildings. Tenants are often unaware that their "security locking system" is subject to the building and fire codes, and some security companies are either ignorant of the fact or choose to violate the code by installing locking systems outside of the construction permit process.

NOTE

Tenants are often unaware that their "security locking system" is subject to the building and fire codes.

The installation of any device on a rated door assembly poses several problems. The obvious problem is that the function of the lock or device conflicts with the intent of the code (see Figure 11-4). All means of egress doors must be openable from the side from which egress is to be made without the use of a key or special knowledge or effort. That means one motion, on one knob, handle, paddle, or crash bar. Adding another locking arrangement may require occupants to perform two operations such as push on a paddle and then turn a knob. This is not acceptable.

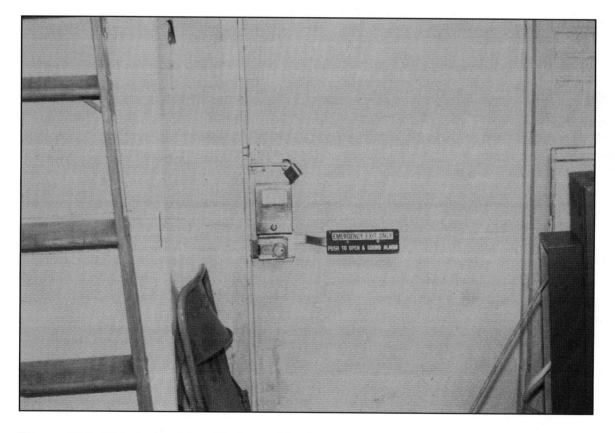

Figure 11-4. This bolt and padlock combination prevent the use of the panic hardware. The installation may also have voided the listing of the fire door and frame.

In addition to the operational problems add-on devices may pose, they may violate the listing of the rated door assembly. Unless the door and add-on device are listed for installation together, the add-on cannot be installed. If it has already been installed, the owner has real problems. NFPA, *Standard for Fire Doors and Windows* provides guidance for the installation of accessories on fire doors, as well as for their repair.

> **NOTE**
>
> **In addition to the operational problems that add-on devices may pose, they may violate the listing of the rated door assembly.**

A permit is required for the installation of delayed egress locks, access-controlled egress doors, and other special locking arrangements. When these are installed without benefit of permit and approval, they may not comply with the code and can potentially pose a significant hazard. For each system there is a list of specific requirements that must be met in order to make use of the special provisions. In each case all of the provisions must be met. Use of the provisions is not a pick-and-choose exercise.

Compromises of the Integrity of the Means of Egress Function

If posed the question "What are the functions of the building areas that comprise the three different parts of the means of egress?" most of us would answer in the same way, "To get people out of the building." We would be right, at least partly. The function of the exit access, exit, and exit discharge together is to get the people out and away from the building. The problem is that only one part of the means of egress, the exit, actually has a part of the building dedicated to it and absolutely nothing else (see Figure 11-5).

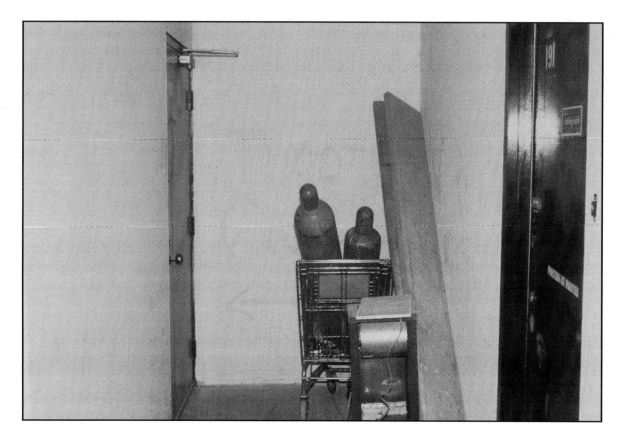

Figure 11-5. No storage whatsoever is permitted in an exit, be it a stair or a passageway in a covered mall.

The exit access is the portion of the building that leads to the exit and to where you work, play, eat, pray, shop, or do anything else people do in buildings. Exit access is merely a path through the usable part of the building. The width of this path varies depending on the occupant load, but this is clearly the most dangerous part of the means of egress. If there is a fire it will probably be in this space, occupants will be exposed while in the exit access.

Exit discharge is the path from the termination of the exit to the public way, and often the part that building owners have the least control over. It is no less important than the other two and is more likely to be obstructed than used for other purposes. Like the exit access, the exit discharge is merely a dedicated path through a general area, not a dedicated building element like the exit. Trash containers, delivered goods, and the belongings of evicted residents are only a few of the enemies of the exit discharge.

NOTE

Trash containers, delivered goods, and the belongings of evicted residents are only a few of the enemies of the exit discharge.

The exit is the only one of the three parts with a specific portion of the building dedicated solely for its use. It may be of considerable size as in the case of a high-rise or covered mall building, or may be only the thickness of a door frame. When exits are used for any purpose other than the movement of people, a dangerous condition has been created. Storage that is combustible or that could become an obstruction can quickly make the exit unusable by the occupants.

Removal or Elimination of Exits

If the building permit process was used every time an addition was built or a temporary structure such as a tent was erected, there would be very few eliminated exits. Unfortunately, some people persist in trying to circumvent the process and what some would say is worse, it is usually the fire inspector who catches it—at the worst possible time.

> **NOTE**
>
> **When exits are used for any purpose other than the movement of people, a dangerous condition has been created.**

Sometimes exits are eliminated when additions are added. What was once an exterior door in a restaurant now leads to a lovely glass-enclosed dining room with a balcony and no exterior stairs. If the work has been performed without a construction permit, there could be serious problems. Since an exit has been removed and the floor area has increased, exit capacity for the new occupant load is certainly questionable. Exit remoteness and travel distance could also have been adversely affected, not to mention the structural integrity of the balcony.

Another problem arises when occupants are forced to exit through an illegally constructed area that may not be structurally sound, or when illegal construction has encroached upon an exit path making it a hazardous area. Assembly occupancies are famous for enclosing outside dining areas for the winter months. If the provisions of the building code and the construction permit process are followed, a safe, serviceable addition to the main structure may be possible. Unfortunately, prefabricated temporary structures are sometimes installed by contractors from outside the jurisdiction who have little regard for the code, permit process, or the manufacturer's installation instructions.

When these temporary structures are encountered by a fire inspector and the determination is made that the structural integrity has not been assessed by the building department, the fire inspector should immediately issue an order that the addition be vacated. Any exits from the existing structure that empty into the addition should be closed off and the building occupant load adjusted accordingly, until the building inspector gives the okay to reopen the new addition. The possibility of injury to the occupants from structural collapse in an uninspected temporary structure is real and necessitates immediate action.

Buildings undergoing renovation during partial occupancy pose a particular challenge to fire and building inspectors and require a close working relationship between the two departments. Provisions to ensure the adequacy and integrity of required exits should be carefully outlined in preconstruction meetings for each phase of the project. Occupants cannot be permitted to pass through construction areas while exiting the building. Any work within the means of egress during business hours should be strictly prohibited.

NOTE

Occupants cannot be permitted to pass through construction areas while exiting the building; any work within the means of egress during business hours should be strictly prohibited.

Exit discharges are especially threatened during construction due to the constant influx of construction materials, equipment, and trailers. Buildings undergoing repair or renovation with partial occupancy require frequent spot checks on a regular basis so that fire prevention and occupant safety remain a priority of the general contractor.

NOTE

Exit discharges are especially threatened during construction due to the constant influx of construction materials, equipment, and trailers.

SPECIAL REQUIREMENTS FOR BUILDINGS AND OCCUPANCIES

Assembly Occupancies

Assembly occupancies have historically suffered the biggest loss of life due to problems with the means of egress. Many of our code requirements can be traced directly back to catastrophic fires in which a high loss of life led to a public demand for action. Many of the large fires we used in our discussions of interior finish and fire-resistive construction also had serious deficiencies in means of egress. The Coconut Grove and Beverly Hills Supper Club are two examples; the Happy Land Social Club in New York City, where eighty-seven people were killed in 1990, was a more recent example.

NOTE

Assembly occupancies have historically suffered the biggest loss of life due to problems with the means of egress.

Deficiencies in the means of egress were not the only problems in these buildings. They did ensure, however, that the occupants were exposed to the deadly effects of fire for an excessive and, in many cases, fatal period of time. Because of our history in assembly occupancies, specific provisions for them have been included in the fire prevention codes.

Posted Occupant Loads and Seating and Egress Plans

All public assemblies for fifty or more persons must have the approved occupant load posted. This is a key item to be checked during routine inspections. The worst time for an inspector to have to try to determine an occupant load is during a performance or during prime time at restaurants or nightclubs.

Overcrowding is not a subjective term. The codes define overcrowding as exceeding the occupant load calculated in accordance with the building code. The fire prevention codes require the occupant load be posted and enforced. There should be no need for discussion between the business owner and inspector as to whether an unsafe condition exists. The code requires specific action by the fire official to ensure that the overcrowding condition is remedied by the business owner. It is the owner or manager who stops the show and takes the necessary steps to come within the posted occupant load. The inspector merely makes him do it.

> **NOTE**
>
> **Overcrowding is not a subjective term; the codes define it as exceeding the occupant load calculated in accordance with the building code.**

Overcrowding conditions, especially involving establishments that serve alcohol, can be particularly difficult and even dangerous for fire officials. Personal safety must be a primary consideration. If police assistance is needed it should be requested, and if necessary the inspector should leave the premises until police assistance arrives. Meeting the business owner in his or her office, where there are fewer distractions has the advantage of allowing the owner to save face and thus be more cooperative, and also keeps potentially belligerent customers out of the picture.

Code language in the IFC and UFC regarding overcrowding has been significantly modified from the legacy model fire codes. Chapter 1 authorizes the fire official to remedy overcrowding. Means of Egress provisions require occupants to evacuate upon orders from the fire official. Specific language prohibiting the owner or operator from allowing overcrowding is missing from the code. Some jurisdictions have amended the code to include it, a wise addition that puts responsibility squarely where it belongs-- the owner-operator of the business.

> **NOTE**
>
> **Overcrowding conditions, especially involving establishments that serve alcohol, can be particularly difficult and even dangerous for fire officials; personal safety must be a primary consideration.**

As we discussed in Chapter 7, many spaces have multiple occupant loads posted depending on whether the occupants are standing, seated at tables, or seated in rows of chairs. Having multiple approved seating plans is an excellent way for public assemblies to have maximum flexibility in the use of their space and yet comply with the code.

> **NOTE**
>
> **Many spaces have multiple occupant loads posted depending on whether the occupants are standing, seated at tables, or seated in rows of chairs.**

Emergency Preparedness

The model fire codes address the issue of emergency preparedness by mandating the development of emergency plans, requiring personnel training in emergency evacuation and proper use of fire extinguishers, and requiring drills in high-risk occupancies. These facilities have large occupant loads, house hazardous processes, or contain persons whose capacity for self-preservation is reduced due to age, medical condition, or confinement. The fire codes also require accurate record keeping on the part of facility managers.

The IFC and UFC contain requirements for appropriately trained "crowd managers" at large events, and for fire evacuation plans, fire safety plans, emergency evacuation drills, and employee training and response procedures. These provisions ensure prompt, appropriate actions on the part of building occupants in any emergency situation necessitating evacuation. Evacuation is only part of the equation. Accurate accountability for employees and occupants after evacuation is also critical.

> **NOTE**
>
> **Fire drills should be unannounced and should include scenarios that test the fire plan.**

An evaluation of emergency plans, drills, and training records should be a part of each routine inspection. Plans and records must be routinely checked, or they will not be maintained by business owners.

Educational Occupancies

Fire drills, considered a nuisance by some educators, are an important part of every school's fire safety plan. Many fire safety professionals do not give school fire drills the attention that they deserve. Drills should be unannounced and should include scenarios that test the fire plan. An administrator might be posted at a particular corridor to inform teachers that for the purposes of the drill, the corridor is full of smoke and impassable. Administrators should also select a few students to remain in the office to see if they are reported as missing.

The time spent witnessing fire drills is productive time for fire inspectors and fire department operations units. The students, teachers, administrators, and parents realize that safety at the school is high on your list of priorities. Face-to-face meetings between operational units and school officials are also the best way to ensure adequate fire department access. Assembly points where the children gather after evacuating the school must not be in areas through which responding fire equipment must pass.

> **NOTE**
>
> **Assembly points where children gather after evacuating a school must not be in areas through which responding fire equipment must pass.**

Records of all drills are required by the model fire prevention codes. As with all records required by the code, a review by the inspector during each routine inspection is mandatory.

Institutional Occupancies

Emergency plans and evacuation procedures for institutional occupancies must be based on the needs of the occupants and the design of the facility. Hospital or nursing home patients who are nonambulatory, or not capable of self-preservation, have very different needs than those in a halfway house. Both may be institutional occupancies, but the safeguards required by the building codes based on the different institutional use groups are quite different.

> **NOTE**
>
> **Emergency plans and evacuation procedures for institutional occupancies must be based on the needs of the occupants and the design of the facility.**

Every institutional emergency plan must address two key elements. The plan must adequately address the needs and limitations of the occupants, and be compatible with the construction features of the building. The emergency plan prepared by a facility can be an unintentional notification to the fire official of an illegal change in use. Occupancies that house persons who are incapable of self-preservation have much stricter construction requirements including fire-resistive construction, smoke partitions, and fire detection and fire sprinkler systems. To safely use the "defend in place" strategy in institutional occupancies, they must be constructed with all of the safeguards required by the building code. Approval of an emergency plan in which nonambulatory residents are instructed to remain in a building without the safety features prescribed by the model building codes could lead to disaster.

NOTE

To safely use the "defend in place" strategy in institutional occupancies, they must be constructed with all of the safeguards required by the building code. The emergency plan prepared by a facility serve as an unintentional notification to the fire official of an illegal change in use.

Participation in drills by fire department operations units is an invaluable test of the compatibility of the fire department preplan and emergency plan of the facility, especially in jails and prisons where firefighter safety takes on an added dimension.

High-Rise Buildings

Fire safety and evacuation plans are required by all the fire prevention codes for high-rise buildings. Like all emergency plans, they should be "exercised" with the fire department. Fire drills are not required by the model codes for high-rise buildings, but they are a good idea. Advance planning and notification of all the building occupants by building management are required because most high-rise fire alarm systems selectively alarm specified floors and areas. Failure to make the necessary notifications can lead to confusion or fear for which the fire department will be blamed, regardless of fault.

Summary

The means of egress is a clear and unobstructed path that leads from any point within a building or structure to a dedicated public way or open space. It is comprised of three distinct elements called the exit access, the exit, and the exit discharge. Exit access and the exit discharge are merely paths, passages, or corridors through the otherwise occupied portions of the building. The exit, however, is a space solely dedicated to the movement of the occupants. It may be expansive as in the case of exit stairways in a high-rise building or be no larger than the thickness of the door frame on a small building.

The exit access and exit discharge must be maintained in a clear and open condition with the appropriate width and height as determined by the building code. The exit must remain completely free of any storage or materials and is dedicated solely to the movement of the building occupants. There are five general ways in which the means of egress may be compromised: Exits are obstructed or obscured, the integrity of their construction is compromised, the integrity of their function is compromised, or the exit is eliminated.

Review Questions

1. List the three elements of a means of egress.

    a. _____
    b. _____
    c. _____

2. List five ways in which the means of egress can be compromised.

    a. _____
    b. _____
    c. _____
    d. _____
    e. _____

3. List two ways in which the means of egress can be obstructed.

    a. _____
    b. _____

4. List two ways in which the means of egress can be obscured:

    a. _____
    b. _____

5. List two ways in which the integrity of construction elements and special features of the means of egress can be compromised.

    a. _____
    b. _____

6. List two ways in which the integrity of the means of egress function can be compromised.

    a. _____
    b. _____

7. List two instances in which exits may be eliminated or removed from service due to safety concerns.

    a. _____
    b. _____

8. Which occupancy classification has historically suffered the largest life loss due to problems with the means of egress?

    _____

9. When an overcrowding condition is encountered by the fire inspector, who is responsible for taking the necessary steps to comply with the approved occupant load?

    _____

10. All the model fire prevention codes require that employees of public assemblies receive training in _____.

Discussion Question

1. You have been directed by the chief's office to look into a complaint by a city council member that a new house of worship is dangerously overcrowded on Friday afternoons. More than 300 cars have been counted in the undersized parking lot and within the surrounding community. The facility is an existing building formerly used as a school and is located within a residential neighborhood. Upon research you discover that the occupant load for the building based on square footage and capacity of the means of egress should be approximately five hundred persons. However, a zoning restriction imposed on the occupancy permit limits the facility to two hundred persons, based on the availability of only one hundred parking spaces in the parking lot.

    a. If on inspection, the facility in fact does have 500 persons in attendance, is it overcrowded?
    b. What occupant load should the fire official enforce?
    c. What would be your report to the chief's office?

KEY TERMS

overcrowding

a condition in which the number of occupants within a building or space exceeds the approved occupant load calculated in accordance with the building code

Chapter 12

Hazardous Materials

Learning Objectives

Upon completion of this chapter, you should be able to:

- Describe the control area concept of hazardous materials management.
- List the steps that must be taken to establish control areas.
- Describe the importance of hazardous materials management plans and hazardous materials inventory statements to the inspections process.
- Describe the NFPA 704, *Standard System for the Identification of the Hazards of Materials for Emergency Response.*

CODE PROVISIONS FOR HAZARDOUS MATERIALS

The 1990s brought new code provisions for the storage, handling, and use of hazardous materials. The four model building and fire codes were updated to better regulate hazardous materials. Some from the old school in fire prevention bureaus thought that the code process was being taken over by hazmat team members. Some even talked of leaving the enforcement of the new code provisions to the fire department hazmat team and only enforcing the *real fire prevention regulations.*

The facts speak differently. The hazardous materials regulated by the reformatted and expanded codes were and are the same hazardous materials that the model fire codes have regulated for years. Hazardous materials are not only things with the words methyl-, ethyl-, or trichloro-, at the beginning of their names. For the purposes of the model building and fire codes, the term *hazardous materials* is a laundry list of substances with potential physical and health hazards. The list runs from combustible liquids, oxidizers and water reactive materials; and from corrosives to toxic and highly toxic solids, liquids, and gases. Some materials regulated by the hazardous materials code provisions aren't actually classified as hazardous materials at all. Cinnamon is a tree bark. The process of grinding and pulverizing cinnamon creates a combustible dust. That process is a high-hazard and must conform to the code.

> **NOTE**
>
> **For the purposes of the model building and fire codes, the term *hazardous materials* is a laundry list of substances with potential physical and health hazards.**

The chapter or article in your fire code entitled Hazardous Materials provides general regulations that apply to all hazardous materials as defined by the code. Material-specific requirements are found within chapters or articles that address specific classes of substances such as flammable and combustible liquids or oxidizers.

The NFPA 704 System

NFPA 704, *Standard System for the Identification of the Hazards of Materials for Emergency Response,* establishes a marking system for buildings and tanks that provides immediate information as to the hazards to health, the flammability, the reactivity (instability), and special hazards of materials (see Figure 12-1). The system uses a diamond divided into four color-coded quadrants. The numbers 0–4 indicate the relative hazards with the number 4 denoting the greatest hazard. The system is also used as a basis for the definition of hazardous production materials (HPM), which is discussed later in this chapter under the section Hazardous Production Material Facilities.

NOTE

NFPA 704 establishes a marking system for buildings and tanks that provides immediate information as to the hazards to health, the flammability, the reactivity (instability), and special hazards of materials.

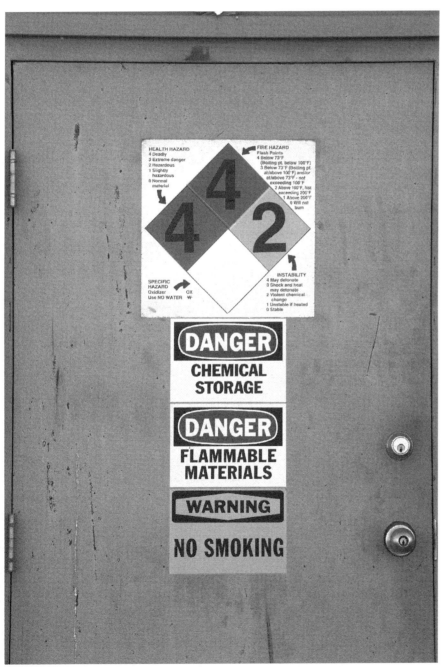

Figure 12-1. The NFPA 704 marking system is mandated by the fire codes for buildings and tanks storing hazardous materials. (© Depositphotos.com/D.P. Fox)

GROUP H OR HIGH-HAZARDS BUILDINGS

In Chapter 3, we learned that Group H buildings are those in which quantities of hazardous materials exceeding the **maximum allowable quantity** (MAQ) are stored, handled, or used. The three legacy model codes identified High Hazard uses based on the amount of regulated material that was stored, handled, or used within the structure. Older editions of the model codes identified High Hazard uses based on process rather than material.

The codes list the materials in tables, by class, state, storage arrangement, and use. For example, there are three classes of flammable liquid, Class IA, IB, and IC. Each has its own MAQ. There is a MAQ for used in an open system such as a dip tank, where vapors are liberated in the air, and a MAQ for closed systems where the liquid is used in sealed equipment and piping. MAQs are typically doubled for sprinklered buildings and doubled for storage in approved storage cabinets. The increases are applied accumulatively.

MATERIAL & CLASS	STATE	MAQ/STATE STORAGE	OPEN/CLOSED SYSTEM MAQs
Class IA Flammable	liquid	30 gallons	Are vapors liberated?

Table 12-1. MAQs are dependent on material, class, state, and use in an open or closed system.

Table 12-2 is a partial listing of Use Group H buildings from the *International Buildings Code* and the later editions of the legacy codes. It's strictly for discussion. Always refer to the specific edition of the code under which the building was designed and constructed. What is different? In the older editions of the model codes, High Hazard uses were identified according to the processes that occurred in the buildings. Ammunition manufacturing was high-hazard because of the process being conducted. By the mid 1980s, the model codes based the classification on types and amounts of material stored, used, or handled in buildings, regardless of the use. Ammunition manufacturing is still high-hazard, but based on the materials NOT the process. If the maximum allowable quantity of a regulated material is exceeded, the building must be designed under the appropriate Group H classification, unless one of the exceptions is applicable.

HAZARD	IBC	BNBC	SBC	UBC
Detonation	H-1	H-1	H-1	H-1
Deflagration	H-2	H-2	H-2	H-2
Physical Hazard	H-3	H-3	H-3	H-3
Health Hazard	H-4	H-4	H-4	H-7
Repair Garage	S-1a	S-1a	S-1a	H-4
Aircraft Hangar	S-1a	S-1a	S-1a	H-5
HPM Facility	H-5	F-1a	F-1a	H-6

Table 12-2. Building code H Use Groups, from the IBC and legacy codes.

Exceptions to the Group H Classification

Prior to conducting an inspection of a Group H occupancy, it is wise to spend some time reviewing the code and records of previous inspections, permits, and other information about the facility. This is good advice prior to conducting every inspection, and is more important when inspecting a Group H building. Most jurisdictions don't have that many Group H occupancies, so a quick code refresher can go a long way.

While most inspectors don't inspect a lot of Group H buildings, they do commonly encounter buildings in which an illegal change in use has occurred, brought on by the amount of regulated material in the building. The typical scenario involves a change in a manufacturing process, where a new chemical or material is introduced. The amount of material needed for the process exceeds the maximum allowable quantity. Three options are available for the owner:

1. Discontinue the new process

2. Upgrade the structure to the appropriate Group H classification

3. Meet one of the exceptions

Table 12-1 described the MAQ system and noted that quantities are typically doubled for sprinkler protection and doubled again for storage in approved storage cabinets. At first glance, the table in the IBC or your legacy code can be a bit unnerving. Are liquor stores, groceries, and hardware stores Group H? Are common household cleaners and products regulated? Doesn't Grandpa take nitroglycerine pills for his angina? Should his house be a Group H? Before you develop chest pains yourself and waste a lot of time digging through the Group H provisions of the code, look at the exceptions first.

The 2012 IBC has 13 exceptions to the group H classification. Generally the following occupancies and processes involving the storage, handling, and use of hazardous materials will not cause the structure to be classified as group H, even if the maximum allowable quantities are exceeded.

- Retail storage of alcoholic beverages in containers not exceeding 1.3 gallons
- Storage of foodstuffs and cosmetics where the quantity of water-miscible flammable liquid is less than 50 percent in containers not exceeding 1 gallon
- Mercantile storage in accordance with the fire codes
- Control areas maintained in accordance with the fire codes

This is not, by any means, a complete listing.

Control Areas

Control areas are areas within buildings that are separated by fire-resistance rated construction and rated opening protectives. The number of control areas that can be established within a structure are regulated by floor. There is also a limit to the number of control areas that are permitted within a building. When building codes first adopted the concept of maximum allowable quantities for hazardous materials, there was a fundamental flaw in the concept: The same maximum quantity applied to a fire-resistance rated building that covered a city block that applied to a 2,500-square-foot frame building on the outskirts of town.

The fundamental flaw in the system was that it applied to all buildings and structures, regardless of the size or construction type. The operator of a factory that covered an entire city block and was constructed of masonry was limited to the maximum allowable quantity (MAQ) of hazardous materials. The 2,500-square-foot wood frame building on the outskirts of town was limited to the exact same amount.

The control area concept allows the building owner to establish multiple areas within a building and to store and use or handle up to the maximum allowable quantity within each one. Although the legacy model codes and the IBC have slightly different provisions, the fundamentals are the same:

- The number of control areas permitted per building and per floor are specified within the building and fire codes.
- The number permitted in mercantile occupancies is limited to two.

- Control areas must be separated from one another by fire separation assemblies with a minimum 1 hour fire-resistance rating.
- Ratings for floors and ceilings are also required.

NOTE

The control area concept allows the building owner to establish multiple control areas within a building and to store and use or handle up to the maximum allowable quantity within each one.

In Figure 12-2, four control areas for the storage of Class II combustible liquid have been established within a factory–industrial building. The maximum allowable quantity of Class II liquids listed in Table 12-1 is 120 gallons. Without the establishment of control areas, the factory would be limited to 120 gallons.

The codes would permit the establishment of four control areas in this one-story building, each containing up to the MAQ. By separating the storage areas with fire barriers, separate control areas are established and the factory can store up to 480 gallons of Class II combustible liquid.

Figure 12-2. This building has 4 control areas, each has the MAQ.

NOTE

The establishment of control areas that involves any construction whatsoever requires the owner to secure a building permit.

Fire inspectors must remember that the establishment of control areas that involves any construction whatsoever requires the owner to secure a building permit. The fire inspector should direct the owner to the building department, so that construction plans can be reviewed, permits secured, and the installation can be inspected.

Another key item is that the number of control areas is regulated per building, not per tenant. Multiple tenants within single buildings are collectively limited to the same number of control areas as a single tenant who occupies the same space. A one-story building is limited to four control areas no matter how many tenants occupy the building. Determining who has property rights to the control areas is not a role nor is it a responsibility of the fire inspector. It is the responsibility of the building owner.

NOTE

The number of control areas is regulated per building, not per tenant. A one-story building is limited to four control areas no matter how many tenants occupy the building.

Control Areas in Mercantile Occupancies

Control areas within mercantile occupancies follow slightly different rules. Individual packages offered for retail sale are generally quite small. Most consumers do not pick up 55-gallon drums of corrosive materials, drop them into a shopping cart with the paper towels and mixed nuts, and stroll over to the cash register. In recognition of this fact, the codes permit mercantile occupancies to exceed the exempt amounts for storage per control area for certain materials, providing certain conditions are met.

These maximum quantities or "mercantile exempt amounts for storage and display," are significantly higher than the base exempt amounts. This exception applies only to the storage and display of nonflammable solid and nonflammable or noncombustible liquids. All other substances are limited to the normal exempt amounts. Additionally, individual containers and packages are limited in size, and displays are limited both in height and by the number of pounds or gallons per square foot. Increased aisle widths are also required.

> **NOTE**
>
> **In addition to the exempt amount for each class of hazardous material, the fire prevention codes have established a permit threshold, and the two are not interchangeable.**

INSPECTIONS OF BUILDINGS WITH HAZARDOUS MATERIALS

A quick look at the building code MAQ tables should bring about the realization that most buildings will contain at least some hazardous materials. The model fire codes contain specific provisions aimed at the management of the inspections process. In addition to the maximum allowable quantity for each class of hazardous material, the fire codes have established a permit threshold. The two are not interchangeable. The permit threshold establishes a level at which inspections for compliance with the code is considered prudent by the collective bodies of the model code organizations. There is no need to justify your inspection. As we discussed in Chapter 2, the model code process has justified it for you.

By using the permit system, you prioritize your inspections and simplify issues surrounding your right of entry. In dealing with hazardous materials code requirements, you also gain valuable information within the permit application process. The model fire codes permit fire official to require the submission of a **Hazardous Materials Inventory Statement** (HMIS) and **Hazardous Materials Management Plan** (HMMP). The model fire codes include sample plans in their appendices. Copies of **Material Safety Data Sheets** (MSDS), which are submitted as a part of the HMIS, are also required to be available on site. Requirements for maintenance vary by code depending on the quantities of hazardous materials. Locking repository containers in approved locations are required in some occupancies.

A thorough and effective inspection of many occupancies is practically impossible without a review of the facility HMIS and HMMP. Imagine attempting to inspect a large warehouse-style hardware store. An inspector would be hard pressed to determine if the occupancy was within the exempt amounts for combustible liquids alone, much less for all of the hazardous materials. Think of all of the products that are classified as combustible liquids—paints, varnishes, cleaners, adhesives, waterproofing, pesticides, and many more. Consider that they are spread throughout the store on shelves and in the back storage areas. Now imagine going through this exercise for each class of regulated material! A review of the HMIS, and perhaps a spot check of selected materials for accuracy of the HMIS, makes the goal of a thorough, complete inspection possible.

Fire officials should also consider requiring the use of pounds, gallons, and cubic feet for reporting quantities in HMIS. SARA Title III, Tier II reports list quantities in pounds for all substances regardless of the physical state. Since the model building and fire codes only use weight for solid materials, inspectors are forced to spend precious time translating pounds to gallons for liquids, or pounds to cubic feet for compressed gases, depending on the substance.

Material Classification and Identification

A methodical, step-by-step approach must be used in determining what code requirements are applicable to a particular substance or product. Many products contain constituents that cause them to fall within several categories. Such materials must comply with the provisions for each category. Liquid pesticides that have a combustible liquid base must comply with the provisions for pesticides and combustible liquids. In determining conformance with the exempt amount/maximum allowable quantity restrictions, the pesticide would be included within the total quantity for toxic or highly toxic materials and the total quantity for combustible liquids.

The first step in the process is a review of the MSDS. The NFPA 704 classification gives general guidance. Do not be surprised if the material poses three or more distinct hazards, each with its own exempt amount/maximum allowable quantity and permit threshold. Use the following process to determine the code requirements for a product:

1. Use the MSDS to determine the different hazards posed by the product (Class 2 Oxidizer, Flammable Liquid, Toxic Materials, etc.)
2. Using the fire prevention code, determine the permit threshold and the exempt quantity for each hazard classification.
3. Apply the general provisions of the article or chapter on hazardous materials.
4. Apply the material specific requirements from the applicable chapters or articles of the code.

NOTE

A thorough and effective inspection of many occupancies is practically impossible without a review of the facility HMIS and HMMP.

Example

Chemical products are used in modern building ventilation systems to kill microorganisms that can grow within the system and be distributed throughout the building. One product in common use is classed as a Class 2 Oxidizer, Corrosive and Toxic Material, and is manufactured in the form of hockey-puck-size tablets.

> **NOTE**
>
> **Many products contain constituents that cause them to fall within several categories, and such materials must comply with the provisions for each category.**

Under the provisions of the IFC, a permit is required to store, use, or handle over 100 pounds of this product in any building or structure. Storage within a single control area is limited to 250 pounds. Keep in mind, however, that this example assumes that this is the only hazardous material within the building. To determine the need for permits and maximum quantity per control area, the total quantity of all materials that fall within a given category would have to be calculated. If a building owner stored 50 pounds of the product from our example and more than 50 pounds of another Class 2 Oxidizer, he would still need a permit. The permit is per class of material, not per specific product.

Storage and handling of the material must also comply with the provisions of all three classifications. Storage of our example product in the basement of a building is prohibited by all model codes under the provisions for Class 2 solid oxidizers. All the requirements for toxic materials and corrosives would also apply.

Hazardous Production Material Facilities

Hazardous production materials (HPM) have a health, flammability, or reactivity of Class 3 or 4 when rated in accordance with NFPA 704. HPM facilities are research, laboratory, or production facilities that use HPM and produce an end product that is not hazardous. Semiconductor manufacturing plants are HPM facilities. The construction requirements, materials storage, and handling requirements are different from those for like materials in other Group H facilities. There is no margin of error in the manufacturing of semiconductor chips, so industrial processes are among the most stringent. HPM facilities quite literally have their own set of rules.

Summary

Code provisions for the storage, use, and handling of hazardous materials have been coordinated between the model building and fire prevention codes to provide alternatives to the High Hazard uses of the past. Materials that pose multiple hazards must comply with the code provisions for each hazard. Permit thresholds and exempt amount/maximum allowable quantity are applied on the basis of total quantity per category, not total quantity per product. The general provisions for all hazardous materials as well as the material specific requirements from throughout the code must all be applied.

Review Questions

1. Which standard establishes a marking system for buildings and tanks that provides information as to the health hazards, flammability, and reactivity of materials? _____

2. The maximum quantity of a regulated substance that can be stored within a control area in a building that is not classified as H or High Hazard is _____.

3. Areas within a building in which hazardous materials in quantities not exceeding the exempt amount may be stored, handled, or used are called _____.

4. The model building and fire prevention codes specify the maximum number of control areas per _____.

5. List two documents that must be submitted with each hazardous material permit application.

    a. _____
    b. _____

6. A _____ or document that describes the relative hazards must be on site for each hazardous material which is used, handled, or stored.

7. Are exempt quantities and permit or approval thresholds for hazardous materials the same? _____

8. List the three types of facilities that may be classified as HPM facilities.

    a. _____
    b. _____
    c. _____

9. Materials that pose multiple hazards must comply with the requirements for _____.

10. Permit or approval thresholds for hazardous materials that pose multiple hazards are based on _____.

Discussion Questions

1. Based on a recommendation from the clerical staff, the new fire marshal is considering dropping the requirement that all permit applications for hazardous materials include a Hazardous Materials Inventory Statement. Complaints of excess paperwork have been received from industry and the clerical staff.

    a. How will this potentially affect the inspections process?
    b. What is your recommendation to the fire marshal?

2. During an inspection of a nonsprinklered factory, you NOTE that 750 pounds of a Class 2 oxidizing material are being stored in the building. The exempt amount is 250 pounds. The manager informs you that he cannot purchase the material from his supplier in quantities less than 500 pounds. The storage area is tidy and in good order and there are no other code violations in the building. What storage options are available to the factory manager under the provisions of the code?

KEY TERMS

maximum allowable quantity (MAQ)

threshold quantity of a hazardous material established by the building and fire codes as the maximum amount that can be stored, handled, or used within a building that is not classified as Use Group H

control areas

area within a building in which hazardous materials in quantities not exceeding the maximum allowable quantity may be stored, handled, or used

Hazardous Materials Inventory Statement (HMIS)

an inventory of regulated materials in a form approved by the fire official that includes the following information: chemical name; manufacturer's name; hazardous ingredients; UN, NA, or CIS identification number; and maximum quantities stored or used on-site at any time

Hazardous Materials Management Plan (HMMP)

a management and contingency plan in a form approved by the fire official that includes the following information: site plan, floor plan, material compatibility, monitoring and security methods, hazard identification, inspection procedures, employee training, and emergency equipment available

Material Safety Data Sheet (MSDS)

a document, prepared in accordance with DOL 29 CFR, that contains information regarding the physical and health hazards associated with a given product or substance and a recommended emergency action

hazardous production material (HPM)

a solid, liquid, or gas that is classified as a 3 or 4 in accordance with NFPA 704 for hazards to health, flammability, or reactivity and used in research or production where the end product is not hazardous

Chapter 13

Flammable Liquids and Aerosols

Learning Objectives

Upon completion of this chapter, you should be able to:

- Describe the system used to classify flammable and combustible liquids.
- Describe the hazards involved in the storage of aerosol products and the protection methods required by the model codes.
- Discuss the importance of using tanks in accordance with their listings.
- Discuss the importance of labeling requirements for containers.

INTRODUCTION

Burning liquids have been with us for a long time. Asphalt was used for roads and to waterproof boats in Mesopotamia prior to 3000 b.c. The ancient Egyptians used it in the mummification process to ensure that Amenhotep, Tutankhamen, and the other pharaohs passed into the afterlife in good condition. The Byzantines perfected the use of combustible liquid as a weapon in their defense of Constantinople in the seventh century a.d. Greek fire, a mixture of crude oil, charcoal, sulfur, and other chemicals, was poured over the surface of the sea and ignited.

Today we drive our gas-guzzling cars on asphalt roads, heat our homes and businesses with fuel oil, and lubricate nearly all machinery with some type of combustible liquid. An average American drinks 2 gallons of beer and wine each year,[1] our food is prepared with combustible vegetable oils, and we spray or rub cosmetics containing alcohol on our bodies.

Flammable liquids and **combustible liquids** are very much a part of our everyday lives. Trips to the local service station that include the operation of self-serve fuel dispensers are a regular part of our routine. Between 1994 and 1998, one in every thirteen service stations in the United States experienced a fire. Two-thirds of the fires involved vehicles and resulted in 7.7 million dollars in property damage. Gasoline was the first material ignited in two-thirds of the vehicle fires. Static electrical discharge and smoking were among the causes.[2]

Physical Properties

What is it about liquids that burn that makes them so dangerous? The very same properties that have made them so useful to us. Regardless of the source, most share certain characteristics. They are excellent fuels, the **heat of combustion** of kerosene is more than double that of wood.[3] They are plentiful, occurring naturally, as in petroleum, or through industrial processes or fermentation. Some such as ethyl alcohol are **water-miscible**, whereas others are not and can be used as water repellents and sealants. They give off flammable vapors at comparatively low temperatures. The **flash point** of gasoline for instance is about −45°F.[4] Flammable and combustible liquids generally ignite at relatively low temperatures, making them readily available sources of heat energy, which is used to power engines, heat boilers and furnaces, and generate electricity. Imagine life without cars, cocktails, or hair spray! Not to mention heat and artificial lighting.

Many code requirements are based on the physical properties of the liquids. Liquids with flash points below normal ambient temperatures are naturally a greater hazard than those that must be heated in order for them to give off flammable vapors. A classification system based on physical characteristics is used by the model codes. Examples are shown in Table 13-1. Requirements for storage, handling, and transportation of flammable and combustible liquids are based on these classifications.

Classification	Flash Point	Boiling Point	Examples[a]
Class IA Flammable	< 73°F	< 100°F	Ethyl ether
Class IB Flammable	< 73°F	≥ 100°F	Ethyl alcohol, gasoline
Class IC Flammable	≥ 73°F and < 100°F	N/A	Butyl ether
Class II Combustible	≥ 100°F	N/A	Fuel oil #1 (kerosene)
Class IIIA Combustible	≥ 140°F and < 200°F	N/A	Fuel oil #6
Class IIIB Combustible	≥ 200°F	N/A	Lubricating oil (motor oil)

Table 13-1. Liquid classifications.

[a] James H. Meidl, *Flammable Hazardous Materials* (Beverly Hills, CA: Glencoe Press, 1970), p. 44.

CODE PROVISIONS

Requirements for the storage, handling, and use of flammable and combustible liquids are found within the model fire codes and referenced standards. NFPA 30, *Flammable and Combustible Liquids Code*, NFPA 30A, *Automotive and Marine Service Station Code*, and NFPA 30B, *Manufacture and Storage of Aerosol Products* are referenced by the model fire codes. Several other standards relating to tank vehicles, oil burning equipment, and repair garages are also referenced. Information for the installation and maintenance of oil burning equipment is also found in the model mechanical codes.

> **NOTE**
>
> **Flammable and combustible liquids generally ignite at relatively low temperatures, making them readily available sources of heat energy.**

General Fire Safety Requirements for Liquids

The misuse and improper handling of flammable and combustible liquids have resulted in staggering losses of lives and property. A fire incident involving a gasoline tanker that overturns on the interstate gets a fair amount of play in the press. The hapless citizen who suffers severe burns from trying to clean paint brushes with gasoline usually does not even make the local paper. You can bet that the paint brush cleaning incidents vastly outnumber the tanker crashes. The model fire codes contain general fire safety requirements in an attempt to protect the public from their own actions and those of others. General safety requirements in all the model codes include:

- All containers (see Figure 13-1), vessels, and tanks containing flammable or combustible liquids must be approved for the purpose and marked or labeled with an appropriate warning.
- All equipment, piping, and fittings must be listed and approved for the specific application, including dispensing apparatus, pumps, and valves.
- Dispensing into unapproved containers is prohibited.
- The sale of Class I liquids as domestic cleaning solvents is expressly prohibited. Class I and II liquids may not be used indoors for parts cleaning, except within machinery listed for that purpose.
- Flammable and combustible liquids must be dispensed in an approved manner. Flammable liquids must be drawn from tanks by pump. Combustible liquids may be dispensed by gravity through approved equipment.
- Flammable and combustible liquids may not be discharged onto the ground or into any waterway. All waste must be disposed of in an approved manner.

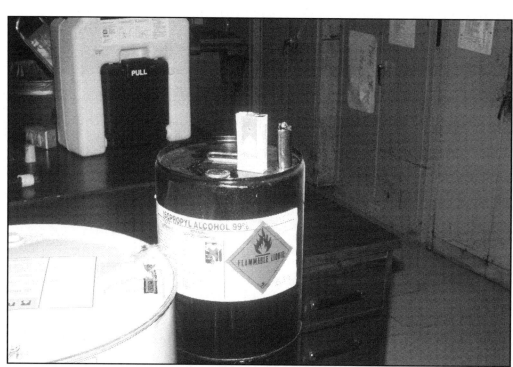

Figure 13-1. Approved containers must be labeled in accordance with DOT requirements.

STORAGE

Containers and Portable Tanks

Storage requirements are based on the type of vessel and classification of the liquid. Code provisions for containers and portable tanks address **containers** up to 119 gallons and portable tanks with up to 660 gallon capacity (see Table 13-2). All containers and portable tanks must be tested and listed for the intended service. Standards for the design of metal drums, safety cans, and plastic containers have been developed by the Department of Transportation (DOT), the American National Standards Institute (ANSI), the American Society for Testing and Materials (ASTM), and Underwriters Laboratories (UL). The appropriate standards are referenced in your model fire code. Whether the storage arrangement is code compliant requires a two-part test: 1) is the container or tank listed and 2) for the liquid in question. NFPA 30 contains a complete list of approved container sizes by type and size, ranging from glass containers to drums and portable tanks. Approved 1-quart glass container s are acceptable for Class 1B liquids. Those same containers are not acceptable for Class 1A liquids, which are limited to 1 pint.

NOTE

All containers and portable tanks must be tested and listed for the intended service.

	FLAMMABLE			COMBUSTIBLE	
TYPE	CLASS IA	CLASS 1B	CLASS 1C	CLASS II	CLASS 111
GLASS	1 pt	1 qt	1.3 gal	1.3 gal	5.3 gal
SAFETY CANS	2.6 gal	5.3 gal	5.3 gal	5.3 gal	5.3 gal
METAL DRUMS	119 gal	119 gal	119 gal	119 gal	119 gal

Table 13-2. Approved container sizes. Source: NFPA 30, 2015 edition.

Quantity limitations within occupancies are based on the use group of the building, class liquid, and storage arrangements including type of sprinkler protection and floor level in the building. As we discussed in Chapter 6, special sprinkler requirements apply to the storage of flammable and combustible liquids, separate from those in NFPA 13. These requirements are incorporated by reference by the model fire codes from NFPA 30, *Flammable and Combustible Liquids Code*.

NOTE

Special sprinkler requirements apply to the storage of flammable and combustible liquids, separate from those in NFPA 13.

Aboveground Storage Tanks

NFPA 30 includes siting requirements for aboveground tanks. The distance tanks can be located from property lines depends on the size of the tank and exposure protection. All tanks, whether aboveground (**AST**) or underground (**UST**), must conform to one of the standards referenced by NFPA 30 shown in Table 13-3.

TANK CONSTRUCTION	STANDARD
Fiberglass reinforced plastic and polyester	UL 1316
Steel	API 12B, 12D, 12F, API 656, UL 58, 80, ANSI UL 142

Table 13-3 Tank design standards.

As an inspector, you will encounter tanks with capacities that range from 275 gallons to several million gallons. Do not think that bigger necessarily means more challenging in gaining code compliance. The manager of the tank farm probably understands that code compliance is a necessary part of doing business. Mr. Wizard, the local garage owner, who has invented his own revolutionary system of doing oil changes by dispensing motor oil with air pressure from a cannibalized home heating oil tank, is probably a bigger challenge. Mr. Wizard has violated the requirement of using tanks and equipment in accordance with their listings. The home heating oil tank is the lightest service tank made. He has pressurized a vessel that was not designed to be pressurized, risking catastrophic rupture. Add to the rupture possibility the fact that pressurized air, with 21 percent oxygen, will be forcing a combustible liquid out of the ruptured tank. Hopefully, you will find Mr. Wizard's invention before your friends on the engine company have to respond to his garage for the fire.

Types of Aboveground Storage Tanks

ASTs range in size from just over 60 gallons to several million gallons. By definition, any vessel with a capacity of 60 gallons or less is a container, although containers can reach 119 gallons in size. The difference between the two is intended use. Containers are used to transport or store. Liquids in tanks are considered to be in use. Large tanks at storage facilities include floating roof tanks that may be open or covered, cone roof tanks, dome roof tanks, and horizontal tanks.

ASTs at most other facilities include unprotected tanks up to 660 gallons used for the storage of fuel oil connected to the building services, portable tanks of up to 660 gallons (see Figure 13-2), tanks installed in special enclosures in accordance with NFPA 30A, and fire-resistance rated tanks (see Figure 13-3).

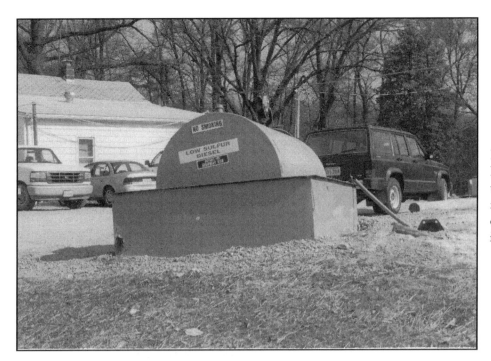

Figure 13-2. This tank has been placed on unlevel ground, rendering the secondary containment inadequate.

Inspection of Aboveground Storage Tanks

Whether you are inspecting a tank farm with a dozen internal floating roof tanks that have an aggregate capacity of 20 million gallons or a 6,000-gallon concrete-encased fire-resistant AST, your inspection involves checking for compliance with the same basic requirements. Small ASTs of ≤ 6,000 gallon capacity are becoming a popular replacement for USTs for the storage of motor fuels and fuel oil for building services. Many building owners are unaware that permits and inspections are required for their installation. Some contractors are also unaware of the requirements or choose to ignore them. Inspectors should pay particular attention when encountering new ASTs. Was the installation permitted and approved? A secondary consideration must also be what happened to the tank that was replaced. Was it abandoned underground also without a permit and in violation of the code?

- Hazard identification is required for stationary ASTs using the NFPA 704 system. Signs do not have to be mounted directly on tanks, but must be readily visible from the outside of diked areas. Physical protection for ASTs and piping from vehicular damage is required. The model fire codes set a performance standard for bollards or barriers. They must be able to resist a force of 12,000 pounds applied 30 inches above the ground surface. They also prescribe guard posts of not less than 4-inch diameter steel filled with concrete, buried to a depth of at least 3 feet, and extending 3 feet above the ground surface.

- Supports, anchorage, and foundations that support tanks must be stable and not subject to undermining from environmental forces nor subject to vehicular damage.
- Dikes around tanks must be maintained free of vegetation and in good repair. Dikes are designed to impound liquid spills due to overfill or tank failure and are designed to contain a spill equal to the capacity of the largest tank within the diked area. Dikes are an engineered fire protection feature. Any changes to the diking system should be designed by a licensed engineer and reviewed by the fire official.
- Vents must be clear and unobstructed and must not terminate in an area that would pose a hazard. Construction that occurs after the installation and approval of an AST can create problems when construction encroaches on vents.
- Control of spillage and housekeeping are key items that protect against damage to the environment and accidental ignition.
- Records for the required maintenance and testing of fixed fire protection systems, fire detection and alarm systems, and fire protection equipment should be checked to verify compliance with the testing requirements of the applicable referenced standards.
- Inventory records should be checked to verify that inventory control practices are in place and are followed by facility personnel. As with all records required by the model fire prevention codes, they must be checked by fire officials during each routine inspection. Failure to ensure that records are complete and in an acceptable form will lead to lax or no record keeping at the facility.

NOTE

Physical protection for ASTs and piping from vehicular damage is required. Changes to diking systems should be designed by a licensed engineer and reviewed by the fire official.

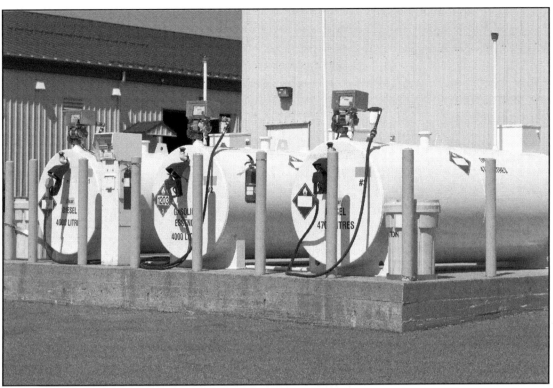

Figure 13-3. The outer layer of these tanks provides a fire resistance rating as well as secondary containment. In spite of the tidiness of the site, the waste containers need to be moved.

Underground Storage Tanks

USTs must be installed in accordance with the requirements of the model codes, applicable referenced standards, and the manufacturer's installation specifications. USTs are tested at the factory and should be tested with air pressure at 3–5 psi, prior to being placed in the excavation. Once the tank is in place and secured to any required tie downs, a third air pressure test must be performed in the presence of the fire official, prior to backfilling. These tests ensure the tank was not damaged in transport or damaged by the contractor during installation. The three tests are complementary and none should be waived by the fire official. They provide assurance to the public and to the owner of the facility that the UST was tight at installation. Piping must also be tested with air pressure for tightness in accordance with your model fire code.

On a rare occasion, an inspector will encounter a tank that was installed and filled without the required tightness tests or inspections. Under no circumstances, should an air test be performed on a tank that contains a flammable or combustible liquid, or that has contained a flammable or combustible liquid. Imagine several thousand cubic feet of compressed air forcing flammable vapors out of a small opening in a tank. Think of the static electrical charge building as the fuel–air mixture rushes out of the tank. Imagine the contractor's explanation after the accident. "I just did what the fire inspector ordered."

Procedures for testing tanks that have contained flammable and combustible liquids are contained in standards referenced by the model codes and should be followed to ensure safety. The contractor should not be released from the requirement that a tightness test be performed. He should be required to submit for approval an alternate test procedure that meets the intent of the tightness test and complies with good engineering practice. In this way, the owner and the public have assurance that the tank is tight and the contractor learns his lesson by having to enlist the service of a consulting engineer or testing contractor.

NOTE

Under no circumstances should an air test be performed on a tank that contains a flammable or combustible liquid.

Routine UST Inspection

Routine UST inspection involves a review of inventory control records and procedures, a visual inspection of exposed vent and vapor recovery piping, and assessment of general fire prevention and spill control practices. Tanks that are out of service must be closed by removal or abandonment in place. Tank **closure** procedures should be strictly observed to guard against underground petroleum releases. Petroleum in the groundwater is a significant fire, health, and environmental hazard.

NOTE

Tank closure procedures should be strictly observed to guard against underground petroleum releases.

Service Stations and Garages

In addition to compliance with code requirements for tank storage, service stations and garages must comply with specific code requirements relating to the dispensing of motor fuels and repair of vehicles. The model fire prevention codes differ in their approach to organizing the requirements, although the basic requirements are very similar.

Supervision is generally required at all service stations open to the public. Attendants must be capable of responding to emergencies at the site, operating emergency shutoff controls and fire protection equipment, and handling routine spills. The attendant is also responsible for preventing the illegal dispensing of Class I liquids into unapproved containers.

> **NOTE**
>
> **The service station attendant is also responsible for preventing the illegal dispensing of Class I liquids into unapproved containers.**

Dispensers

Dispensers, commonly referred to as "pumps," should be carefully inspected. Public exposure to the hazards associated with flammable and combustible liquids is probably the greatest at the dispenser island of their local service station.

- Dispenser protection must be provided by installing barrier protection or raising the dispensers on a 6-inch island. Lots that have been repaved, raising the surface to within 6 inches of the top of dispenser islands must be restored to the original level or have barrier protection installed, or the islands must be raised.
- Hoses, nozzles, and fittings must be listed and in good condition. Removal of automatic hold-open levers by station operators in violation of the listing voids the listing on the nozzle.
- Emergency shutoff features that control power to the pumps and the flow of fuel should be inspected. A remote master shutoff that controls electric power to the pumps must be accessible and prominently marked. An approved emergency shutoff valve, which automatically closes in the event of fire or impact, must be installed in each dispenser. Valves must be tested at the time of installation and be retested by the operator annually.

> **NOTE**
>
> **Public exposure to the hazards associated with flammable and combustible liquids is probably the greatest at the dispenser island of their local service station.**

AEROSOL STORAGE

In 1982, a fire at the Kmart distribution center in Bucks County, Pennsylvania, demonstrated the potential hazards of the storage of **aerosol** products. The 1.2-million-square-foot facility was fully sprinklered. Storage was palletized and on racks and was 15 feet high. In-rack sprinklers were not provided.

A case of aerosol carburetor cleaner was knocked off a pallet and fell to the floor. An employee operating a propane powered forklift reported hearing a hissing sound following immediately by flames. He attempted to reach a fire extinguisher, but was driven back by the fire, which had extended up the face of the stack from which the case had originally fallen. Other employees attempted to extinguish the fire but were also forced to evacuate.

The fire, fueled by the forty to fifty pallets of carburetor cleaner in the vicinity of the point of ignition broke through the roof within 13 minutes. Ruptured aerosol containers rocketed through the structure, trailing burning contents and rapidly spreading the fire. The fire burned out of control for 8½ hours and was not totally extinguished for 8 days. Property damage alone exceeded $100 million, and worse, Kmart's distribution system had been destroyed.[5]

Full-scale fire tests conducted during the late 1970s and early 1980s demonstrated the potential for rapid fire development associated with the rupture and rocketing of aerosol containers.[6] In many of the tests, conventional sprinkler protection was quickly overwhelmed. Standard storage arrangements including aisle widths and travel distances were inadequate. As a result, specific requirements for the storage of aerosols were incorporated in the model fire prevention and building codes and NFPA 30B, Code for the Manufacture and Storage of Aerosol Products was developed.

Aerosol Classifications

Aerosol products are classified Level 1, 2, or 3, based on the chemical heat of combustion of the base product and the propellant. Storage requirements are based on this classification. Each carton of aerosol product must be clearly marked on at least one side with the words "Level ___ Aerosols." Unmarked cartons are automatically considered to be Level 3, or the most hazardous. Products such as whipped cream, in which neither the base product nor the propellant are flammable, are not classified and are not regulated under these code sections.

> **NOTE**
>
> **Each carton of aerosol product must be clearly marked on at least one side with the words "Level ___ Aerosols."**

Storage Requirements

There are five basic storage configurations for aerosol products. Each has specific requirements for quantity limits, storage configuration, and fire protection.

1. Aerosol or liquid warehouses are special structures specifically used for the storage of flammable liquids and aerosol products. The IFC and UFC use the term aerosol warehouse.
2. General purpose warehouses are buildings where various commodities are stored within a single building. Distribution centers for hardware and supermarket

chains as well as other mercantile occupancies are general purpose warehouses. For the purposes of aerosol storage requirements, storage areas in mercantile buildings are also general purpose warehouses. Storage arrangements as well as quantities often depend on the season. Deck stain and patio furniture do not sell well in Minnesota in January. Two storage arrangements are permitted in general purpose warehouses. **Nonsegregated storage**, in which aerosol products are stored throughout the facility intermixed with other commodities, is regulated depending on storage arrangement (palletized or rack storage). Quantities permitted within **segregated storage** areas are significantly higher. Segregated storage areas are enclosed by rated walls or chain-link fence enclosures and are protected with sprinkler systems designed to apply water at much higher densities than required by NFPA 13.

3. Retail sales areas are limited to specific quantities depending on the sprinkler protection provided. Permitted quantities are increased if the area is sprinklered. An additional increase is given if the sprinkler system is designed to protect areas used for the storage of aerosols.
4. Storage in A, B, E, F, I, and R use buildings is limited to 1,000 pounds of Level 2; 500 pounds of Level 3; or 1,000 pounds of Levels 2 and 3 combined.
5. Outdoor storage of aerosol products, including products stored in trailers or temporary buildings, must be located with respect to exposures, including buildings and lot lines, and exit discharges.

NOTE

Segregated storage areas are enclosed by rated walls or chain-link fence enclosures.

APPLICATION OF FLAMMABLE FINISHES

Flammable finishes are applied to vehicles, furniture, and various other consumer items through spraying or dipping operations. The result is a durable finish that is hard to match using nonflammable products. The use of low flash point liquids results in rapid drying and ease of multiple applications. Low flash point liquids, especially when atomized during spray finishing operations, are particularly unforgiving.

NOTE

Low flash point liquids, especially when atomized during spray finishing operations, are particularly unforgiving.

Spray Finishing

Spray finishing must be conducted in an area designed and constructed for that purpose. Spray finishing is specifically prohibited in A, E, I, or R occupancy buildings, unless within a properly constructed spray room. Spray finishing is only permitted within approved **spray booths**, **spray rooms**, or **spray areas**.

At auto body repair facilities, employees sometimes spray outside spray booths during particularly busy times. This is strictly prohibited and extremely dangerous. Power sanding, grinding, and other spark-producing activities provide a ready ignition source. Inspectors should question signs of overspray on garage floors and caution against spraying in unauthorized areas.

NOTE

Inspectors should question signs of overspray on garage floors and caution against spraying in unauthorized areas.

Inspections of spray finish facilities should include attention to the following conditions: Paint and solvent storage must be in approved cabinets or in a room constructed for that purpose. Liquids must be in approved safety cans or approved containers. Quantities must comply with the code requirements. Approved self-closing waste cans (see Figure 13-4) must be provided. Flammable liquid containers must be bonded to prevent against static discharges. Fire doors that protect paint storage rooms are commonly propped open or the closer assemblies are removed. These rated opening protectives must be maintained in accordance with the code.

Figure 13-4. Approved rag cans must be provided and used. (Courtesy of Ron Berry)

Approved fire extinguishers should be visible, accessible, and maintained in accordance with NFPA 10. Particular attention should be paid to portable fire extinguishing equipment in auto body shops—it gets used!

> **NOTE**
>
> **Particular attention should be paid to portable fire extinguishing equipment in auto body shops—it gets used!**

Spray booths must be separated from combustible storage or construction by a 3-foot clear area, or a rated wall or partition. Filters must be clean and permit the designed air flow through the ventilation system. Discarded filters must be immediately removed to a safe detached location or stored submerged in water until discarded. Booths should be kept clean and free of residue, and sprinkler heads should be protected with thin plastic bags that are frequently changed.

Spray rooms must be maintained in a clean condition. Combustible storage within the spray room is not permitted. Inspectors should ensure that filters are clean and that the ventilation system is operating efficiently. Residue buildup on ventilating fans quickly destroys their efficiency. All ventilation systems must discharge to a safe location, away from ignition sources, other air intakes, building openings, or exit discharges.

Dipping Operations

Dip tanks, into which items are submerged to apply flammable finishes, must be installed within areas that comply with building code requirements. Depending on the size of the tank, an overflow system, emergency bottom drain, automatic closing cover, and fixed fire suppression may be required. Open flames, spark-producing equipment, or unapproved electrical equipment are prohibited within the vapor area.

Inspectors should ensure that approved self-closing waste cans are provided and that waste material is disposed of properly. Suppression systems should be maintained in accordance with appropriate standards and their listings. Ventilation equipment must operate efficiently and discharge to safe locations.

Automobile Undercoating

The spray application of automobile undercoating is permitted to be conducted outside of approved spray booths or spray rooms only if adequate natural or mechanical ventilation is provided and the undercoating materials have a flash point $\geq 100°F$.

Dry Cleaning

Code requirements for the construction, operation, and maintenance of dry cleaning plants depend on the class solvent used in the dry cleaning process. Dry cleaning establishments range from pick-up and delivery sites to self-service facilities that use nonflammable solvents, to dry cleaning plants that use combustible solvents. Building code provisions for use group classification and special occupancy requirements, as well as the special requirements of the fire prevention codes, are based on solvent classification as shown in Table 13-4:

CLASS SOLVENT	FLASH POINT
I	< 100°F
II	≥ 100°F <140°F
IIIA	≥ 140°F and < 200°F
IIIB	≥ 200°F
IV	Nonflammable

Table 13-4. Dry cleaning solvents and flash points.

The model fire codes prohibit the change of solvents to another classification without prior approval by the fire official. Perhaps as important, the machinery in use must be listed for the solvent. All tanks, piping, and equipment must also be listed for the intended use.

NOTE

The model fire codes prohibit the change of solvents to another classification without prior approval by the fire official.

Summary

The same physical characteristics that make flammable and combustible liquids useful make them treacherous. They emit flammable vapors at low temperatures, often well below normal ambient temperature. They have high heat of combustion ratings, often double that of wood and common combustibles.

The hazards associated with flammable and combustible liquids are managed by the codes by regulating specific areas:

- All equipment, containers, tanks, and piping must be listed for the intended use. This minimizes fires and accidents resulting from equipment and container failure.
- Quantities of liquids are limited according to building use and fire protection features.
- Ignition potential is minimized by prohibiting open flames, spark-producing processes, and unapproved electrical equipment.

Review Questions

1. What two physical properties of flammable and combustible liquids are used to classify them? a. _____ b. _____

2. The amount of heat given off by a particular substance during the combustion process is called _____.

3. The minimum temperature at which a liquid gives off sufficient vapor to form an ignitable mixture at the surface but not sufficient to sustain combustion is called its _____.

4. Name the standard developed by the National Fire Protection Association to regulate the manufacture, storage, and handling of aerosol products. _____

5. _____ liquids may be dispensed by gravity from approved equipment.

6. Special sprinkler requirements for flammable liquid storage facilities can be found in _____.

7. A container is any vessel with a capacity equal to or less than _____.

8. Stationary tanks must be identified using the _____ hazard identification system.

9. Underground storage tanks (USTs) are tested with air at _____ psi prior to being placed within the excavation.

10. The storage arrangement in which aerosol products are stored throughout the facility intermixed with other commodities is called _____ storage.

Discussion Question

1. During an inspection of a service station, you observe a customer dispensing one dollar's worth of gasoline into a plastic antifreeze jug. He explains that he has run out of gas and his car is two blocks away. The attendant told him that without a twenty dollar deposit he could not borrow the metal safety can the station maintains for just this reason. On questioning, the attendant verifies the customer's statements. The station has lost three safety cans in four weeks to customers who have run out of gas. Besides, the antifreeze jug is listed for this use since antifreeze is combustible.

a. Should you permit the customer to proceed to his car with the gasoline?
b. What violation(s) of the code have occurred?
c. Is the attendant right about the antifreeze jug's listing?
d. What would be your actions

KEY TERMS

flammable liquid

a liquid having a flash point below 100°F and further categorized based on flash point and boiling point as types IA, IB, and IC

combustible liquid

a liquid having a flash point at or above 100°F and further categorized based on flash point as types II, IIIA, and IIIB

heat of combustion

the amount of heat given off by a particular substance during the combustion process; a measure of fuel efficiency

water-miscible

water soluble

flash point

the minimum temperature at which a liquid gives off sufficient vapor to form an ignitable mixture at the surface, but not sufficient to sustain combustion

container

a vessel with a capacity of 119 gallons or less used for the storage or transportation of flammable or combustible liquids (piping and engine fuel tanks are not containers)

AST

aboveground storage tank for regulated liquids

UST

underground storage tank for regulated liquids

closure

tank closure; placing a tank permanently out of service by removal or abandonment in place using an approved method

aerosol

a product that is dispensed from a container by means of a liquefied or compressed gas

nonsegregated storage

the storage of aerosol products in general purpose warehouses within areas that are not used exclusively for the storage of aerosols

segregated storage

the storage of aerosol products within areas specifically designed for that use incorporating code-required storage arrangements and fire protection features

spray booth

a structure designed and constructed to be used for the application of flammable finishes, featuring power ventilation, fixed fire suppression, and separation that is installed within a building

spray room

a room designed and constructed for the application of flammable finishes complying with the requirements of the building code

spray area

an area designed and constructed to be used for the application of flammable finishes complying with the requirements of the building code

David Diamantes

Chapter 14

Detonation and Deflagration Hazards

Learning Objectives

Upon completion of this chapter, you should be able to:

- List and describe the three basic categories of explosive materials in commercial use.
- Describe two different methods for the initiation of explosive materials.
- Describe the duties and responsibilities of the fire official in allegations of property damage or accidents involving commercial blasting.
- Describe the duties and responsibilities of the fire official in permitting public fireworks displays.
- Describe the handling of misfires of explosive materials and fireworks.

EXPLOSIVES

Depending on whose book you read, the first explosive material was developed by the Chinese, Arabs, or Byzantines. A monk named Roger Bacon first wrote the formula for black powder in the thirteenth century. Another monk, Berthold Schwartz, put Bacon's formula to use and invented the gun a century later. He was remembered by the epitaph: *"Here lies Berthold the Black, the most abominable of inhumans, who by his invention has brought misery to the rest of humanity."* And so, black powder got its name not from the monk who discovered it, but from a monk named Berthold, who developed a practical use for the discovery. I'd hardly count Berthold as an abominable inhuman.

NOTE

Black powder has been described as "treacherous," because its ignition temperature is so low that the slightest spark will ignite it at normal temperatures.

By the seventeenth century, black powder was used in mining throughout Europe. The first recorded use of black powder in Colonial America was in 1773 when a copper mine in Connecticut was converted into the Newgate Prison. In 1802, Eleuthere duPont deNemours began the commercial production of black powder near Wilmington, Delaware. Eleuthere lost his life in an explosion at the mill but members of his family continued production at what eventually became the chemical manufacturing giant DuPont.

Black powder has been described as "treacherous," because its ignition temperature is so low that the slightest spark will ignite it at normal temperatures.[1] Black powder **deflagrates**, or burns very rapidly but at subsonic speeds. Black powder opened up the Midwest by blasting through the Appalachian and Rocky mountains so that the railroads could cross. Fatal accidents were a continual reminder of the unpredictability of black powder. A man and a mule were said to have been lost to accidents every day in crossing the Appalachians.

Nitroglycerine was discovered in 1846 by an Italian scientist named Ascanio Sobrero who was so appalled by its destructive power that he cautioned against its industrial use.[2] In 1862, Alfred Nobel took over his father's company, which manufactured a nitroglycerine headache remedy called Glonoin Oil.[3] He began testing nitroglycerine for use in blasting, and although his brother was killed during the testing, he refused to give up. By 1866, he had discovered a method of using diatomaceous earth to absorb nitroglycerine, thus making it reasonably stable. He named his invention dynamite after the Greek *dynamis* or power.

In addition to the ease of handling and relative stability of Nobel's invention, a key safety factor was the means of initiation. Unlike black powder, a static spark or even a burning fuse will not initiate dynamite. It takes a small explosion to start the explosive reaction. Early **detonators** were gun (percussion) caps loaded with fulminate of mercury, which were fired by means of a lit fuse. Today detonators are manufactured with such precision that delay blasting, in which multiple charges are fired at the same time and initiate explosive reactions milliseconds apart, is the norm.

NOTE

Unlike black powder, a static spark or even a burning fuse will not initiate dynamite.

EXPLOSIVE MATERIALS

The 1900s saw a tremendous increase in the use of explosives in the United States. The United States consumed 287 million pounds of black powder in 1907; United States underground explosives consumption reached 100 million pounds in 1957, and 7.4 billion pounds of explosives were used in the U.S. in 2012.[4]

Commercial explosives are used in four major industries: coal mines 54% of commercial blasting; construction and miscellaneous blasting 16%; metal mines 15%; and nonmetallic mines and quarries 15%. The mining, quarrying, and construction industries use over 4 billion pounds of commercial explosives per year in the United States to facilitate rock excavation activities.[5]

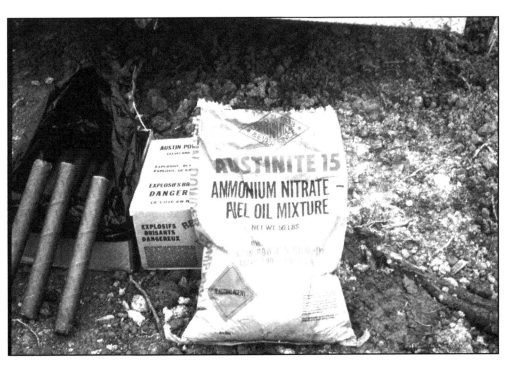

Figure 14-1. Commercial explosives.

Explosive materials (see Figure 14-1) is a catchall term that includes explosives, blasting agents, black powder, detonators, or blasting caps and other products, some of which are simply formulations of the preceding list. The relative safety of these materials varies greatly, as do the requirements for storage and handling. **Explosives** storage is limited to **magazines** with strict requirements for construction. **Blasting agents** are commonly stored in the same 40-foot semitrailer that transported them to the site. The erection of a fence and a padlock for the trailer are often the only required precautions.

> **NOTE**
>
> **Blasting agents are commonly stored in the same 40-foot semitrailer that transported them to the site.**

The role of the fire inspector in regulating the storage, use, and transportation of explosives is normally limited to the areas of construction blasting, demolition, and special industrial explosives use. Mining and quarrying operations, by far the largest users of commercial explosive materials, are generally regulated by state government mining authorities. Interstate and inter-jurisdictional transportation of explosive materials falls under the authority of the Department of Transportation and state police agencies.

Permits are required by the fire codes for the manufacturing, use, sales, display, possession, disposal, or transportation of explosive materials. Some states and some jurisdictions have established blaster certification programs similar to those established by state mine regulators. Blasters are tested and issued licenses or certificates of competency. Background checks have become the norm. Where tests have been properly developed and are technically valid, such programs are often embraced by industry as well as the public as a means of ensuring competent blasters.

Unfortunately, competency testing generally ensures the minimum acceptable level of performance. Explosives are particularly unforgiving, and minimal competence is never enough. Effective regulation forces the industry to establish internal procedures to ensure compliance, as well as promoting increased technical competence. Lax enforcement or occasional enforcement merely ensure that when an accident does occur, the public and elected officials will rightly ask what provisions were in place to protect them and who dropped the ball. Effective regulation is impossible without a basic knowledge of the process.

> **NOTE**
>
> **Unfortunately, competency testing generally ensures the minimum acceptable level of performance.**

BLASTING THEORY

Commercial blasting operations (see Figure 14-2) use explosive materials to break up or shatter rock or other material. Generally rock formations are in the way of the road cut, or water or sewer line, or perhaps the basement of a new home. Holes are drilled in the rock to a specified depth in a specific pattern. The drill pattern, as well as the type and amount of explosives are elements of the blast design that must be determined by the blaster. The blast energy from a given amount of explosive material might shatter a hard rock such as limestone. The same amount of explosive energy might be absorbed by a softer material such as shale, and then be transmitted on through the ground to cause property damage. *Less explosive material does not always mean less potential damage.*

Figure 14-2. Holes are drilled into subsurface rock formations and explosives are used to fracture the rock for the excavators.

NOTE

Less explosive material does not always mean less potential damage.

Amount of Explosives per Detonation

The amount of explosive material that is used within a single shot depends on the type and amount of rock or other material to be broken up, the depth of the formation, the amount of dirt or cover that rests on top of the rock formation, and proximity to other structures, roads, or utilities. As early as 1934 when E. H. Rockwell developed his Energy Formula,[5] studies on damage caused by blasting operations have resulted in various formulas that limit charge weight or amount of explosive material detonated per shot.

NFPA 495, *Explosive Materials Code* includes provisions to limit potential damage by limiting the amount of explosives detonated at one time or by monitoring the effects of the blasting with scientific instruments and maintaining blast effects within a safe range.

Scaled distance formulas are the simplest method of controlling potential damage. NFPA 495[6] provides formulas that determine amount of explosives permitted based on the distance to the nearest occupied building. When blasting within 300 feet of the nearest building, the formula $W \text{ (lb)} = (D \text{ (ft)}/50)^2$ is prescribed, where W is the weight of explosives in pounds and D is the distance to the nearest occupied structure in feet. The maximum quantities are per delay interval of 8 milliseconds or more. This means that if charges are set to detonate at least 8 milliseconds apart, they can be considered distinct detonations.

NOTE

Scaled distance formulas are the simplest method of controlling potential damage.

Firing the Shot

The method of initiation or "firing the shot" is critical and depends on the type of detonators used. Detonators, commonly called blasting caps, are inserted into explosives cartridges. The detonators may be either instantaneous or delay types. **Instantaneous type detonators** initiate a single charge per detonator. They may be electric or nonelectric, but the entire load of explosive material is detonated at once.

Delay type detonators usually delay detonation between adjacent boreholes between 10 and 60 milliseconds. This separation of explosive reactions, however slight, enables the blaster to use the same charge weight dependent on distance as used in instantaneous blasting, except per delay instead of per shot. In the examples in Figure 14-3, a blaster has ten holes drilled and is 200 feet from the nearest building. Using the scaled distance formula W (lb) = (D (ft)/50)², he would determine his charge weight dependent on distance as follows:

$$W = (200/50)^2$$
$$= 4^2$$
$$= 16$$

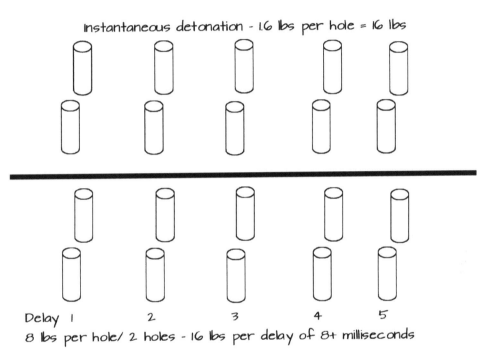

The blaster would be limited to 16 pounds of explosive material per delay of 8 milliseconds or more meaning 1.6 pounds per hole using instantaneous detonators where all ten holes fire simultaneously. Using the same example except with delay type detonators, the blaster could use up to 16 pounds per delay or 8 pounds per borehole because only two fire simultaneously. The benefits of delay blasting are obvious. Blasters can get the job done with fewer shots and with less potential property damage.

NOTE

The benefits of delay blasting are obvious; blasters can get the job done with fewer shots and with less potential property damage.

Blasting Agents

Blasting agents are explosive materials that are not cap sensitive under normal conditions of storage and use. Commonly known by the acronym ANFO (for ammonium nitrate + fuel oil), blasting agents are a relatively cheap, safe, and efficient method of blasting. Because blasting agents are not cap sensitive, they must be used in conjunction with high explosives that are. A high explosive **primer** is used to initiate the blasting agent.

> **NOTE**
>
> **Blasting agents are explosive materials that are not cap sensitive under normal conditions of storage and use.**

Blasters in the field commonly refer to explosives-grade prilled ammonium nitrate as "fertilizer," but do not kid yourself. You would not want to put this stuff on your garden. The use of ammonium nitrate in explosives was patented in Sweden in 1847.[7] One day in 1921, workers at Badische Anilin und Soda Fabrik plant in Oppau, Germany, did what they had done over 16,000 times before when they needed some ammonium nitrate. It was stored outside, exposed to the weather in a 4,500-ton crusted mound. They drilled holes and set explosives charges. Something that day was different. When the smoke cleared, a 50-foot deep crater 450 feet in diameter was all that was left of what was the factory. More than four hundred people were dead and the shock wave was felt 145 miles away.[8]

When determining charge weight per delay, one pound of blasting agent equals one pound of explosives.

> **NOTE**
>
> **When determining charge weight per delay, one pound of blasting agent equals one pound of explosives.**

Blasting Safeguards

The type and amount of cover and proper **stemming** are key elements in reducing the possibility of damage or injury due to **fly-rock**. The first question that must be answered at any blasting operation is where will the rock break to? Rock that is hammered by explosives at supersonic speed and shattered moves. Without a free face or area of relief to break to, the rock will travel in the path of least resistance—up. This is not a big deal on a highway construction job a mile from any development. It is a big deal in populated areas or near utilities, railroads or highways that are open.

NOTE

Without a free face or area of relief to break to, the rock will travel in the path of least resistance—up.

The amount of cover or dirt that covers a rock formation is extremely important. Without adequate earthen cover, rock cannot be safely blasted. Blasting mats (see Figure 14-4), which are used to help confine the effects of the blast, are not a substitute for adequate earthen cover. They are an enhancement, not a replacement.

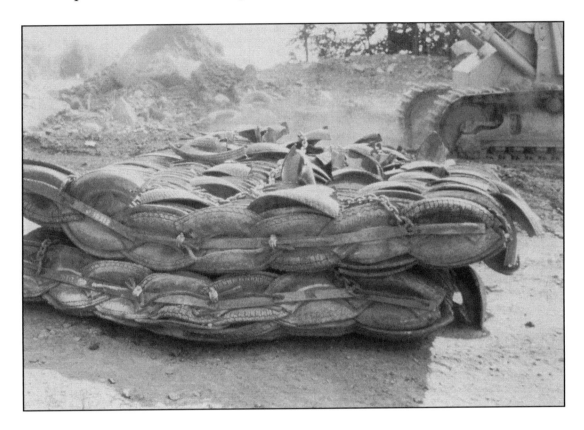

Figure 14-4. Blasting mats used in conjunction with adequate earthen cover, prevent rock from being thrown out of the area.

Without adequate stemming, a borehole is a lot like a loaded mortar. You have a charge at the closed end and an explosive reaction with nowhere to go but out the barrel. Effective stemming confines the effects of the blast and maximizes rock breakage. It also minimizes fly-rock. Rock drill cuttings from the boreholes make poor stemming material. Clean, crushed stone is the most effective material. It must be carefully shoveled into the borehole after loading to avoid any damage to the electrical leg wires of the detonator (see Figure 14-5).

Figure 14-5. The borehole is stemmed by hand in order to protect detonator leg wires from damage.

NOTE

Effective stemming confines the effects of the blast and maximizes rock breakage.

How much stemming is also very important. The formula: T = 0.7 to 1.3 X B

Where T is stemming height and B is the horizontal distance between boreholes is the textbook standard (See Figure 14-6).[9] A rough rule of thumb used in quarries and surface mines is twice the diameter of the borehole in feet. It is a little heavy with the small diameter holes used in construction blasting, but it gets you close.

Proper blast design and adequate stemming maximize the effectiveness of explosive materials and minimize, but do not eliminate, the possibility of fly-rock. There is no substitute for maintaining safe distances from the blast area and for providing adequate shelter for the blaster. In April 1994, a helper on a construction site was struck and fatally injured by fly-rock while standing 236 feet from the nearest borehole. In

December 1999, an equipment operator was struck by fly-rock while seated in the cab of a pickup 800 feet from the blast site. He was guarding the access road to the site.[10]

CONSTRUCTION BLASTING

Shortly after three o'clock on the morning of November 29, 1988, six firefighters were killed in a series of explosions at a construction site. Two fires, one involving a pickup truck and one involving a semitrailer containing blasting agents, were found by arriving crews. While attempting to extinguish the fire in the semitrailer, it exploded, killing the crews of both engine companies. Approximately 40 minutes later, a second trailer exploded, fortunately without additional injuries.[11] Other incidents where firefighters have been killed fighting fires in semitrailers containing explosive materials occurred in 1959 and 1964. Nineteen people, including four firefighters, were lost in the two incidents.[12]

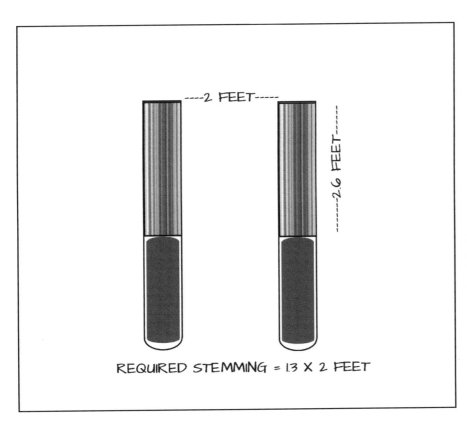

Figure 14-6. The amount of stemming is based on the burden or distance between the boreholes.

Explosive materials are used in road and utility construction across the United States every day. The object is to break the rock into pieces that are easily dug by heavy equipment. Blasting activity depends on the amount and type of underground rock formations within the jurisdiction and the economic pressure to develop. The cost associated with the removal of underground rock may shift development to areas where blasting operations are not necessary. Unfortunately, this sometimes leads to spotty development.

Tracts of vacant, often wooded land that require blasting to put in every sewer line or dig every basement, end up surrounded by established communities. People in the established communities frankly are not particularly interested in any new development. The need for new housing ended abruptly the day their new home was finished. They do not want the construction traffic. They thought that land was park land, after all—and they do not want their house to be damaged by blasting operations. Blasting operations associated with development adjacent to established neighborhoods is one of the most challenging regulatory exercises a fire official will ever experience.

Permits for the storage, use, transportation, sale, disposal, and manufacture of explosives are required under the model fire prevention codes. Permits enable the fire official to keep track of all operations and activities involving explosive materials, and allow the fire official to establish administrative guidelines as conditions of permit.

> **NOTE**
>
> **Blasting operations associated with development adjacent to established neighborhoods is one of the most challenging regulatory exercises a fire official will ever experience.**

Explosives Storage

The storage of explosive materials in any quantity is restricted to approved magazines. Magazine types and construction can be found in each of the model fire prevention codes, NFPA 495, and in the Institute for Makers of Explosives (IME) Safety Publications Library. There are five types of magazines (types 1–5), each with a particular use and limits as to the quantities and types of explosives that may be stored. Requirements range from extensive, for type 1 magazines (weather-, bullet-, and fire-resistant with redundant locking arrangements) to minimal (weather-resistant and lockable) for day boxes.

NOTE

The storage of explosives materials in any quantity is restricted to approved magazines.

Magazine location and storage capacity are determined by compliance with the American Table of Distances approved by IME. The table sets maximum storage amounts in magazines depending on distance from other magazines, inhabited structures, and rail lines and highways. Quantities are increased (significantly at the low end of the table) when the magazine is effectively barricaded. Artificial barricades must be mounds or revetted walls of earth, a minimum of 3 feet thick. Natural barricades are simply the natural features of the area and may be a combination of the terrain and timber such that surrounding exposures cannot be seen from the magazine when the trees are bare of leaves.

Storage within magazines is closely regulated by the fire prevention codes. Explosives and detonators must be stored in different magazines. Storage requirements for high explosives and blasting agents are different. Although blasting agents can be stored in portable trailer type magazines, high explosives cannot. Strict requirements for lighting, locking, and security are contained within the codes.

Problems associated with the storage of explosive materials at construction sites can be a challenge for the fire official. Problems with security and potential fire exposure are common. These conditions are easier to control at permanent storage facilities such as those maintained at mines and quarries. Blasting operations may be the responsibility of a subcontractor who does not control the movement and storage of large amounts of fuel and combustible materials that are stored and used in the operation. Regardless, the blasting contractor is responsible for the safe storage and handling of explosive materials on the site.

NOTE

The blasting contractor is responsible for the safe storage and handling of explosive materials on the site.

If the general contractor is storing combustible materials too close to the magazines, the blasting contractor only has two choices. Get the situation fixed or move the explosives to another site. Inspectors should remember that the role of the fire code official is to ensure the safety of the public through the effective enforcement of the fire code, not the arbitration of disputes between general contractors or subcontractors. Permits for the storage and use of explosive materials should be immediately revoked upon a finding of unsafe or reckless storage use or handling. Second chances are few with explosive materials.

NOTE

Permits for the storage and use of explosive materials should be immediately revoked upon finding of unsafe or reckless storage use or handling.

Explosives Transportation

The inter-jurisdictional transportation of explosives is generally the responsibility of the Department of Transportation and state police agencies. Transportation within individual jurisdictions is addressed by the model fire codes and is serious business. Explosives and detonators can only be transported on the same truck when the procedures of DOT 49 CFR *Specification for the Transportation of Explosive and Other Dangerous Articles* are followed (see Figure 14–7). Detonators and explosives are segregated by hauling the detonators in a steel box lined with 1-inch thick oak.

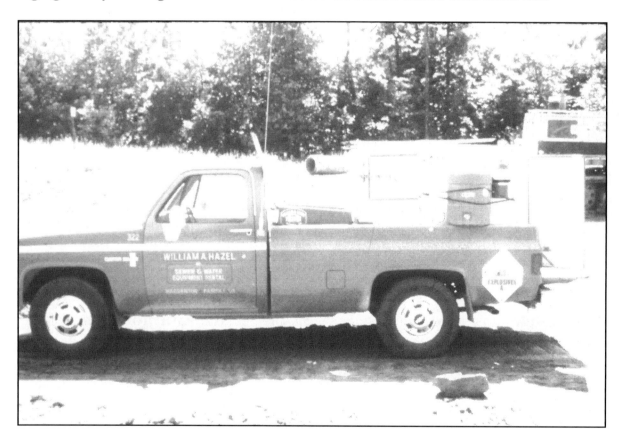

Figure 14-7. Explosives are transported to job sites by truck.

Trucks hauling explosives between magazines and job sites must travel by the least congested routes. Trucks traveling in the same direction must maintain a 300-foot distance between each other. They must be mechanically sound, equipped with fire extinguishers, and not carry unauthorized passengers.

Blasting Operations

Prior to the issuance of a permit for the storage or use of explosives, a thorough inspection of the site should be made. Permit applications should include a map or scale diagram of the site indicating the location of occupied structures, utilities, and highways or rail lines.

Operators of utilities such as water, gas, electricity, and telecommunications must be notified by the blasting contractor at least 24 hours in advance. The utility may decide to have a representative on site during the operation. If blasting is conducted within the utility company's right-of-way, they may have requirements that exceed those of the fire prevention code. In such cases, the blaster must comply with the utility company's wishes, although you cannot enforce the utility company's regulations. You are there to ensure compliance with the fire prevention code.

It is not unreasonable to require contractors to notify the occupants of homes that are in close proximity to the blast area. Blasting complaints can often be minimized if neighboring homeowners know in advance how long the operation will be and are aware that there is government oversight of the blasting operation.

NOTE

It is not unreasonable to require contractors to notify the occupants of homes that are in close proximity to the blast area.

Blast Records

Shot records are not specifically required by the model fire codes, although they are specified in NFPA 495. Nationally recognized good practice dictates that the blaster should have records for each shot. The investigation of any accident or allegation of property damage as a result of blasting will depend on the quantity and quality of information recorded by the blaster. For every shot the following minimum information should be recorded:

- Contractor name, job location, date, and time of day
- Type of rock, number of boreholes, and spacing
- Diameter and depth of boreholes
- Type and amount of explosives
- Amount of explosives per delay of 8 milliseconds or greater
- Method of firing and type of circuit

- Direction and distance to nearest occupied structure
- Weather conditions
- Earthen cover and (or) mats used
- Type of detonators and delay periods
- Type and height of stemming
- Seismograph records (if used)

Good records are the best (and only) defense against allegations of damage by surrounding property owners. Most unfounded allegations of damage are not made by dishonest people seeking to take advantage of the blaster. People often become aware that there is a potential for blast damage and then notice a crack in the concrete basement floor or brick chimney for the first time.

NOTE

Nationally recognized good practice dictates that the blaster should have records for each shot.

NOTE

Good records are the best (and only) defense against allegations of damage by surrounding property owners.

Resolution of these allegations are the responsibility of trained investigators hired by the blaster's insurance carrier. Blast engineers with extensive training and education are able to determine if damage is the result of blasting. The fire inspector's responsibility is to ensure that operations are conducted in accordance with the code. The blaster is probably liable for the damage (depending on state liability laws), whether he was operating within the code or not.

Misfires and Accidents

Misfires and accidents are unusual, but inspectors must be prepared to deal with them. Misfires can occur when the insulation on the leg wire of an electric detonator is skinned, or shifting rock "cuts-off" a hole in the middle of a shot, causing only part of the loaded boreholes to fire. Misfires can also occur if the shot is wired incorrectly or if detonators simply malfunction. Misfires must be handled in a methodical manner with extreme caution:

- The area must be barricaded to ensure that no persons gain access.
- The power source used to fire the shot must be disconnected and electric wires shunted (twisted together) or made safe.

- All personnel should stay out of the area for at least 1 hour if fuse caps are used and ½ hour for other type detonators.
- Only the blaster should then reenter and inspect.
- If possible, the remaining holes should be reconnected and fired.
- If necessary, stemming should be removed by hand, a detonator should be inserted, and the hole fired from the top of the column down.

NOTE

Misfires must be handled in a methodical manner with extreme caution.

If these methods are ineffective, outside help may have to be called in. The role of the fire official in misfires is to see that they are handled safely and expeditiously, and that a minimum number of persons are put at risk. Fire inspectors should never enter the blast area after a misfire.

NOTE

Fire inspectors should never enter the blast area after a misfire.

Accidents involving explosives generally are investigated by state or federal occupational safety, health officials, and law enforcement agencies tasked with explosives investigations. In some jurisdictions, it is the fire prevention bureau or fire marshal's office. The inspector should see that the scene is safe and secured to prevent the destruction of evidence. Like the fire scene, the scene of a blasting accident is often full of activity by emergency personnel. Rescue workers may treat and transport the injured, leaving a path of boot prints and medical supply wrappers. The inspector should make every attempt to prevent unnecessary destruction of the scene.

Other Explosive Materials

Small arms, ammunition, and reloading supplies such as smokeless powder, black sporting powder, and small arms primers are commonly sold to hobby shooters and sportsmen. Industrial explosives and explosives actuated power tools are also commonly found in construction and in industrial applications.

Retail sales and display of smokeless powder (gunpowder), commercially manufactured black sporting powder, and small arms primers are limited. If stored within indoor portable magazines, up to 400 pounds of smokeless powder and 25 pounds of black sporting powder may be permitted to be stored in a fully sprinklered building. Displays are limited to 25 pounds of smokeless powder in original containers. The display of black sporting powder is expressly prohibited.

Mercantile occupancies can and should be guided into compliance with the code. Gun and hobby shows require good communication and a willingness to educate the public. Reloaders and black powder enthusiasts attend these exhibitions, and vendors are only too willing to supply them with materials. Without clear guidelines for the operators of exhibition facilities and regular inspections during the hours of show operations, exhibit hall operators may not police their tenants. Get ahead of the problem. Work with convention center operators to develop a written plan. You may find they have one from another city that you can approve or amend to fit the situation. The important thing is that everyone understands the rules.

NOTE

Without clear guidelines and regular inspections during the hours of show operations, exhibit hall operators may not police their tenants.

Without an adequate plan, all the ingredients for a serious accident or fire are present: a large crowd, quantities of explosive material, and a lack of security. The model fire codes and NFPA 495 adequately address this issue. An adequate program of enforcement by fire officials is needed.

Industrial explosives and explosives actuated tools must be stored and used in accordance with the fire prevention codes. The weight of explosive materials is the determining factor for storage and handling requirements.

Reactive Materials

You will not find provisions for the storage of fertilizer-grade ammonium nitrate in the chapter on explosives in the fire prevention code. Fertilizer-grade ammonium nitrate is not an explosive material even though it is detonatable under the right conditions. Ammonium nitrate and some other substances such as nitromethane and acetaldehyde are classified as unstable (reactive) substances by the codes and are regulated as hazardous materials. Once ammonium nitrate has been formulated into a blasting agent, the provisions of the explosives chapters apply. Until that time, it is an unstable (reactive) substance and regulated as such. Storage in other than Use Group H buildings is limited to the exempt quantities per building or control area as discussed in Chapter 12.

NOTE

Fertilizer-grade ammonium nitrate is not an explosive material even though it is detonatable under the right conditions.

FIREWORKS

Fireworks is a broad term that can include everything from sparklers to multiple-charge aerial shells fired from mortars. DOT 27 CFR classifies common and special fireworks and gives examples of each. An estimated 10,000 persons are injured each year as a result of fireworks injuries. Fire loss and associated medical costs are in the millions of dollars.

> **NOTE**
>
> **An estimated 10,000 persons are injured each year as a result of fireworks injuries, and the fire loss and associated medical costs are in the millions of dollars.**

Why does the American public stand for such a public safety menace? The answer is simple. To the American public, fireworks are every bit as important to the Fourth of July celebration as evergreen trees are to Christmas. There may not be a religious significance attached to the Fourth, but fireworks are a tradition that we hold dear.

Fireworks ordinances generally fall into two categories: those regulating common or consumer fireworks and those regulating aerial and large ground displays. Many jurisdictions have fireworks ordinances separate and distinct from the fire prevention code, which may ban some or all types of common fireworks. Small explosive firecrackers commonly called lady fingers are legal common fireworks according to DOT 27 CFR. Many jurisdictions, however, ban any firework that explodes, acts as a projectile, or rises vertically in the air.

Fireworks Displays

The model fire prevention codes require permits for public fireworks displays, and specifically require the submission of a site plan as part of the permit application. The IFC and UFC reference NFPA 1123 Public Display of Fireworks, which requires a site plan and additional information.

Permits

Permits for public fireworks displays should not be issued without a careful review of the following information:

- Detailed site plan showing all structures, roadways and utilities, spectator and parking areas, and firing and fallout areas

- List and size of all fireworks to be used in the display, manufacturer, supplier, and contact telephone number
- Direction and angle of discharge
- Firing method, mortar size, and construction (see Figure 14-8)
- Name, age, and credentials of pyrotechnician in charge
- Name, age, and credentials of all helpers
- Method of crowd and site access control and number of personnel
- Fireworks arrival date, storage, and security arrangements
- Method of two-way communications between operators and security personnel
- Arrangements for misfires and unspent shells
- Arrangements for range cleanup and inspection immediately after the display and at first light
- Bond or insurance certification

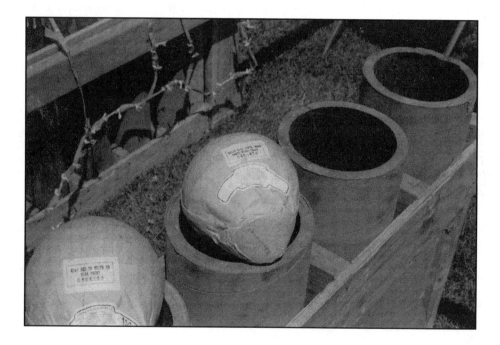

Figure 14.8 Shells must be sorted by size for loading in appropriate mortars.

Minimum clearances dependent on shell diameter are specified in NFPA 1123. The specified distances should be considered the bare minimum, and should never be relaxed. Use of the specified distances will not prevent the fallout of debris and ash outside of the fallout zone. Wind conditions aloft may also be different than conditions at ground level and can carry debris some distance.

NOTE

The specified clearance distances should be considered the bare minimum, and should never be relaxed.

Figure 14-9. Firing areas must be secured to prevent the entry of unauthorized persons.

Firing the Display

Firing the display must be a coordinated effort between the pyrotechnician in charge, his helpers, and site security personnel. The firing area must be secured as shown in Figure 14-9. If at any time lapses in security permit unauthorized persons to enter the fallout area, the fire inspector is obligated to stop the display immediately. The word "shall" in the code indicates that stopping the display is not an option.

NOTE

The word "shall" in the code indicates that stopping the display is not an option.

When shells do not fire, they must be left in the mortar and undisturbed for at least 15 minutes. The mortar should then be flooded with water and left at least 5 more minutes. After the display has ended the shell should be placed in a bucket of water and picked up for transport by the supplier. Malfunctioning shells should never be repaired.

A thorough inspection of the fallout area immediately after the show and again at first light is critical. Live, unfired projectiles are commonly found in the fallout area. In the hands of curious children, these can be deadly. They should be treated as misfires and stored immersed in water for pick up by the supplier.

NOTE

The thorough inspection of the fallout area immediately after the show and then at first light is critical.

Seizure and Destruction of Explosive Materials and Fireworks

The model fire codes provide for the seizure and destruction of explosive materials and fireworks under certain conditions. Confiscation generally gets the attention of even the most uncooperative of code violators. Think before you act. Confiscating fireworks can lead to storage and disposal problems (see Figure 14-10). You and the government are obligated to provide the same safe and code-compliant storage facilities and employ the same safe disposal procedures that the code requires for everyone else. If you seize dynamite illegally stored in the back of a truck parked in a motel parking lot and then illegally store it in a shed at the fire training center, what have you accomplished? Seizure is not always the preferred method of handling a problem.

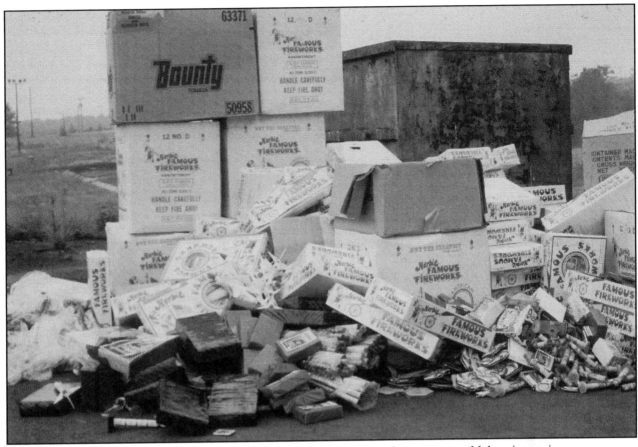

Figure 14-10. Disposal of seized fireworks is time-consuming, dangerous and labor intensive.

Summary

Explosive materials are stored and routinely used in nearly every jurisdiction. Nowhere within the inspection and enforcement process is the inspector's ability to rapidly assess a situation and order appropriate actions more important. Explosive materials are unforgiving and second chances are few.

Inspectors must realize that the difference between an expert in blasting and explosives code enforcement and an expert in blasting and explosives handling is significant. Inspectors should ensure compliance with the fire prevention code and not attempt to serve as operational consultants.

Review Questions

1. Deflagration occurs at _____ speed.
2. Detonation occurs at _____ speed.
3. List four explosive materials.

    a. _____
    b. _____
    c. _____
    d. _____

4. Explosives are stored within structures called _____.
5. A formula used to determine the maximum charge weight per delay of explosive materials dependent on distance, without damage to surrounding structures is called a _____.
6. ANFO, or ammonium nitrate and fuel oil mixtures, are _____.
7. The most effective type of stemming material is _____.
8. At aerial firework displays, fallout areas are established based on shell _____.
9. When aerial firework shells do not fire, they should be left undisturbed for _____ minutes, then _____.

Discussion Question

1. You have been summoned to the chief's office to explain the rash of blasting complaints that have inundated the office staff over the past two days. A large vacant tract formerly used by the youth soccer league for a playing field is being developed into low-cost apartments. Neighboring homeowners complain that the blasting operation is shaking their houses and destroying their foundations.

    a. What code provisions are in place to protect the homeowners?
    b. What code provisions protect the interest of the contractor and developer?
    c. What should be your role in the process?

KEY TERMS

deflagration

very rapid but subsonic oxidation evolving heat, light, and a low-energy pressure wave that is capable of causing damage

detonator

a device, consisting of electric and nonelectric blasting caps, fuse caps, and detonating cord delay connectors, that contains a primary or initiating explosive designed to set off an explosive reaction

explosive materials

explosives, blasting agents, and detonators, including dynamites, slurries, emulsions and water gels, black powder, smokeless powder, detonators and safety fuses, squibs, detonating cord, and other materials whose primary function is to function by explosion

explosives

chemical compounds or mixtures whose primary function is to function by explosion and that cause a sudden and almost instantaneous release of pressure, gas, and heat

magazine

a structure designed and constructed for the storage of explosive materials

blasting agents

materials or mixtures containing fuel and an oxidizer not otherwise classified as an explosive and that cannot be detonated by means of a #8 blasting cap when unconfined

scaled distance formula

a formula used to determine maximum amount of explosive material that can be detonated per delay interval of 8 milliseconds or greater, based on distance to the nearest occupied structure

detonator, instantaneous type

a detonator with no time lapse between the firing signal and detonation of the main explosive charge

detonator, delay type

a detonator that introduces a specific time lapse between the firing signal and detonation of the main explosive charge

primers

explosives packages made up of an explosive charge and a detonator or detonating cord used to initiate other less sensitive explosives or blasting agents

stemming

inert material placed in the borehole after the explosive material has been loaded to confine the effects of the explosive reaction; may also separate charges within a single borehole

fly-rock

rock propelled from the blast area by blasting operations

Chapter 15

Hazardous Assembly Occupancies

Learning Objectives

Upon completion of this chapter, you should be able to:

- Describe the hazards associated with heliports and helistops on buildings, and how the model codes address them.
- Describe the hazards that make airport terminal buildings require special fire protection features.
- Describe the hazards associated with bowling pin and lane refinishing.
- List five general safety provisions for tents and air-supported structures.

INTRODUCTION

A list of fire incidents in which there was large loss of life includes many that occurred where large numbers of people were assembled. Assembly occupancies include those buildings where people gather for civic, social, and religious functions, for entertainment, and awaiting transportation. Fires like the Beverly Hills Supper Club in 1977 with 165 fatalities and the Station Nightclub in 2003 with 100 fatalities immediately come to mind. However, other assembly occupancies are not immune to the effects of fire. There were 168 fatalities in the Ringling Brothers and Barnum and Bailey Circus tent fire in 1944. A gas explosion at the Indiana State Fairgrounds Coliseum in 1963 claimed 74 lives, and in April 1996, a fire at the Dusseldorf Airport terminal killed 17 people.

The model fire codes have specific requirements that address special assembly occupancies such as tents, air terminals, and bowling alleys. These requirements extend the level of protection afforded by the model building codes by regulating the processes and activities within them. Limits on the storage and use of flammable and combustible materials, prohibitions against the use of open flames and heat-producing appliances, and requirements for permits and routine inspections are all included in the model fire prevention codes.

AIR TRANSPORTATION FACILITIES

After the Dusseldorf Airport fire, NFPA's chief fire investigator was quoted as saying "You just don't get structural airport fires like this. You definitely don't get fatalities like this."[1] His comments echoed the feeling of many fire and safety professionals around the world. Millions of people move through airports around the world. Very rarely do fire incidents that do not involve aircraft or aircraft fueling endanger the public.

The model fire codes have operational requirements for airports, **heliports** and **helistops**, and aircraft hangars. The model building codes have construction requirements that pertain to all three. They also include specific requirements relating to helicopter landing areas located on buildings. Unless specifically designed for the purpose, few buildings will accommodate helicopter landings.

NOTE

Unless specifically designed for the purpose, few buildings will accommodate helicopter landings.

Heliports and Helistops

Helicopters take off and land from the ground, from buildings, and from bodies of water at heliports and helistops. Heliports include areas for storage and maintenance and for refueling operations. Helistops are strictly for take-off and landing, with the loading and unloading of passengers or freight.

The increased use of medivac helicopters has led to the establishment of heliports and helistops at many medical facilities. Because takeoff and landing operations account for the majority of helicopter accidents,[2] special building code provisions for heliports and helistops located on structures are crucial to the safety of the occupants. In the case of hospital heliports and helistops, the occupants of the building may be incapable of evacuation. Building code requirements include structural provisions for the landing pad, special exit provisions, and fire protection requirements.

> **NOTE**
>
> **Because takeoff and landing operations account for the majority of helicopter accidents, special buildings code provisions for helicopters and helistops located on structures are crucial to the safety of the occupants.**

Fire code requirements address minimum clearances, fire extinguishing equipment, and general fire safety precautions. Permit requirements in the model fire prevention codes are designed to give the fire official notice of the establishment of these facilities and ensure they are inspected with regular frequency.

Airports

Airports terminals are generally large noncombustible structures. Fuel loads within the passenger terminals are limited. Aircraft service and repair areas are a different story. Fire code requirements regarding passenger safety during loading and unloading and refueling operations specifically address control by airline and airport service personnel. Airline and airport personnel must ensure that no smoking is permitted during these times and must ensure that passengers proceed immediately from the terminal to the aircraft. Refueling with passengers on board is only permitted if airline personnel are on board to ensure that passengers evacuate on the other side of the aircraft in the event of an incident.

Aircraft Hangars

Aircraft hangars are classed by the model building codes as moderate hazard storage facilities. Aircraft hangar fires are often high fire losses due to the value of the aircraft. The large quantities of flammable liquids and solvents and the ignition potential from machinery and tools necessitate strict fire prevention practices. No smoking should be strictly enforced. Fire protection equipment and systems should be scrupulously maintained. Housekeeping should be meticulous.

Bowling Alleys

Two hazardous processes in bowling alleys created the need for special chapters or articles in the model fire codes. Bowling pin and floor refinishing using flammable finishes involves two hazardous conditions. Floor sanding, in which power sanding equipment removes the old floor finish and wood surface creates a large volume of combustible dust. Floor and pin refinishing using flammable finishes creates hazardous vapor–air mixtures. In the case of floor finishing, the entire building is often filled with flammable vapors.

On-site pin refinishing was common when pins were made from solid blocks of maple and before the advent of nylon coatings and injection molding. Pins that are refinished are processed off-site. Floor sanding remains a hazard, particularly if flammable finish is used.

A permit is required by the model fire for floor or refinishing using Class I or II finishes. The permit requirements allow the fire official to inspect the facility during operations to ensure compliance with the code. In addition to general fire safety precautions:

- Floor resurfacing operations are prohibited while the facility is open to the public. This includes both sanding and the application of finishes.
- All heating, cooling, and ventilating systems that recirculate air must be shut down and not operated within 1 hour of application.
- All motors and ignition sources must be shut down and not operated within 1 hour. Smoking is also prohibited.

- Waste material and dust must be stored in approved containers.

The development of high quality latex floor finishes and synthetic bowling pins have greatly reduced the use of flammable finishes in bowling alleys.

TENTS AND AIR SUPPORTED STRUCTURES

The IFC defines a tent as "a structure, enclosure, or shelter without sidewalls or drops, constructed of fabric or pliable material supported by any manner except by air or the contents that it protects." The code has extensive requirements regarding location, anchorage, fabric envelope design and construction, means of egress, and illumination.

The types and amounts of combustible material, heat-producing appliances, smoking and open flames, and flammable and combustible liquids are also regulated. Tents and membrane (air-supported) structures that are to be used for more than 180 days within a 12-month period must also comply with the building code. Certain provisions apply to all tents, canopies, and membrane structures:

- They must be structurally sound and in good repair. Anchoring and inflation systems must be installed in accordance with the building code.
- Fabric coverings must be of flame-retardant material or be treated in an approved manner. Certification must be provided.
- Seating plans must be submitted for approval with construction documents during permit application.
- Open flame devices or unapproved heating devices are prohibited inside or within 20 feet of the tent or air-supported structures.
- Combustible materials such as hay are prohibited except as needed for the feeding of animals. Sawdust or shavings must be kept damp.
- Exit lights and exit illumination are required.
- Heating and cooking equipment must be approved.

- Vegetation and trash or waste material must be kept cleared from around the tent or air-supported structure.
- Approved fire extinguishers must be provided.

Problems regarding the application of the fire prevention code regarding tents can have considerable political impact. Citizens who are unaware of permit requirements or regulations are not normally at their best when a fire inspector arrives during their daughter's wedding reception. The effect can be enhanced when the inspector orders the caterer to stop cooking.

Permit requirements and code provisions should be given to the customer during the permit application process. Tent suppliers within the jurisdiction should be well versed in exactly what the fire prevention bureau requires and expects. The political impact of one inspector's actions can affect the whole fire department during the next budget cycle. I am not advocating lax enforcement. I am advocating no surprises.

NOTE

The political impact of one inspector's actions can affect the whole fire department.

Summary

Fires in special assembly occupancies combine the hazards of high occupant load with the unique fire potential of the specific occupancy. Airports would be less hazardous than large mercantile occupancies (less combustible contents), if not for the aircraft and fuel that surround them. A wedding reception at the Elks Lodge is not particularly hazardous. The same event in a tent can create a multitude of hazardous conditions. The model fire prevention codes require permits for certain processes to enable the fire official to effectively regulate them. Directives can be given at the time of permit application. Inspections will verify that the process or occupancy is in compliance with the code.

Review Questions

1. List three building code requirements for heliports and helistops located on buildings.

    a. _____
    b. _____
    c. _____

2. Under what circumstances are passengers allowed to board an aircraft that is being refueled? _____
3. The major hazard associated with aircraft hangars is _____.
4. What is a major hazard associated with bowling alleys? _____
5. When resurfacing bowling lanes with flammable floor finish, all motors, fans, and ignition sources must be shut down for at least _____.
6. Tent fabric must be made of _____ or be _____ in an approved manner.
7. Under what circumstances are combustible materials such as hay and sawdust permitted inside of tents? _____
8. The majority of helicopter accidents occur during _____.
9. Who is responsible to ensure that no smoking rules are observed during aircraft passenger loading and unloading? _____

10. Open flame devices or unapproved heating devices are prohibited inside or within _____ of a tent or air-supported structure.

Discussion Question

1. You have been contacted by the manager of the local bowling alley. She is refinishing the floors with a latex floor finish starting Saturday night at midnight. She will be sanding and finishing half the lanes Saturday night and reopening on Sunday. She plans on sanding the other half Sunday during the day and will wait to apply the floor finish until Sunday night after closing.

    a. Does she need a permit to conduct this operation?
    b. Does her plan comply with the provisions of your model fire code?

KEY TERMS

heliport

an area where helicopters take off and land from the ground or water, or from a building; includes areas for storage, maintenance, and refueling

helistop

same as heliport, except without facilities for storage, maintenance, or refueling

Chapter 16

Other Storage and Processing Occupancies

Learning Objectives

Upon completion of this chapter, you should be able to:

- Discuss the value of corporate safety programs in industrial occupancies.
- List three of the five main causes and contributing factors involving storage facility fires.
- Explain the reasons for the proliferation of waste material and rubbish handling facilities.
- Describe how the provisions of the model fire prevention codes attempt to reduce fire risk at lumberyards.

INTRODUCTION

NFPA's annual summary on large loss fires during 2011, identified 17 large-loss structure fires, seven of which occurred in manufacturing properties. These included a calcium carbide manufacturing plant, a fertilizer manufacturing plant, an adhesives manufacturing plant, a paper coating plant, a vehicle parts assembly plant, a food processing plant, and a chemical manufacturing plant. These seven fires resulted in total losses of $130.5 million.[1] The actual loss of such fires is much greater than the loss of the building and contents. The loss of jobs and buying power has a catastrophic effect on the entire community, and the effects can be far reaching. Factories across the country can be idled by the abrupt loss of parts from a supplier. In one way or another, these fires affect all of us, either directly or through increased costs for consumer goods, insurance, or consumer credit.

Investigation reports of fires in industrial occupancies often identify factors that contributed to the large loss. Inadequate fire protection system design, improper maintenance, and change in hazard classification are commonly noted, as is the storage of incompatible materials, temporary storage within aisles, and increased height of storage.[2]

All of these issues are addressed by the model fire codes. An effective inspection program that includes the education of facility managers can significantly reduce the likelihood of a large fire loss. Similar fires in chicken processing plants in Hamlet, North Carolina, and North Little Rock, Arkansas, clearly underscore the value of fire prevention practices.

> **NOTE**
>
> **An effective inspection program that includes the education of facility managers can effectively reduce the likelihood of a large fire loss.**

The fires occurred in chicken processing plants, approximately 3 months apart, in 1991. The Tyson plant in North Little Rock had a corporate fire prevention program in place that included the appointment of a fire safety director, mandated fire drills, an established fire brigade, and restricted combustible storage in hazardous areas. The fire, which occurred when a high pressure hydraulic line failed and atomized hydraulic fluid was ignited, caused 8 million dollars in damage and resulted in the plant being closed 13 weeks. All 115 employees evacuated within 3 minutes, and all were accounted for using the procedures practiced during routine fire drills. Fire brigade personnel with self-contained breathing apparatus (SCBA) led suppression crews to the fire area.[3]

In contrast, the fire at Imperial Foods in Hamlet, North Carolina, resulted in twenty-five fatalities and fifty-four injuries among the approximately ninety persons working at the plant at the time of the fire. As in the North Little Rock fire, a ruptured hydraulic line in the cooking area led to the fire. Locked and blocked exit routes and the lack of an evacuation plan led to widespread confusion. One employee perished when she became wedged between a compactor seal and the building wall. Twelve employees took refuge in a walk-in cooler and succumbed to the fire.

Chief Fuller of the Hamlet Fire Department was quoted as saying that the entire incident centered around one problem—the lack of enforcing existing codes. The Hamlet Fire Department had no code enforcement, and the North Carolina Department of Occupational Safety and Health had never inspected the facility during its 11 years in operation.[4]

STORAGE FACILITIES

The main causes or contributing factors in fires in storage occupancies are bad housekeeping, industrial trucks (forklifts), welding and cutting, smoking, and arson.[5] All are addressed by specific fire prevention code provisions.

In Chapter 3 we learned that the model codes contain significantly different requirements for storage facilities, depending on the hazards posed by the commodities stored. Warehouses could be classified as low, moderate, or high hazard, and even within those classifications, NFPA 13 has different sprinkler requirements depending on the commodities stored. Inspections of storage facilities should include verification that the facility has not changed to a more hazardous use. A check of the hydraulic nameplate affixed to the sprinkler riser will indicate the original hazard classification based on the design density (refer to Figure 10-3).

> **NOTE**
>
> **The model codes contain significantly different requirements for storage facilities, depending on the hazards posed by the commodities stored. Inspections of storage facilities should include verification that the facility has not changed to a more hazardous use.**

Idle wooden or plastic pallets create a significant fire load. A maximum storage height of 6 feet is all that ordinary hazard sprinkler protection can effectively protect.[6] Storage heights over 6 feet require an increased density. The outdoor storage of idle pallets must comply with fire requirements regarding pile height, size, distance from structures, and setback from property lines (see Figure 16-1).

Figure 16-1. Idle pallet storage must comply with fire code requirements.

NOTE

Idle wooden or plastic pallets create a significant fire load; a maximum storage height of 6 feet is all that ordinary hazard sprinkler protection can effectively protect.

Warehouse facilities should have designated smoking areas (even if they are outside, they should be designated as such), with suitable ashtrays. Waste containers should be substantial and should be emptied on a regular basis. Spare cylinders for forklifts should be secure, protected against damage, and comply with quantity restrictions of the fire prevention code.

The storage of matches, flammable liquids, aerosols, and other hazardous items within general purpose warehouses entails significant restrictions and safeguards. The code provisions from the applicable chapters or articles of your model fire prevention code and the referenced standards must be enforced.

WASTE MATERIAL HANDLING PLANTS

Developments since the late 1980s have changed the way we feel about trash. When our landfills began to fill up and the public began to support environmental issues, laws were passed to encourage and in some cases mandate recycling. That trash that fire inspectors have battled for years, relegating it to covered metal containers to be hauled quickly away suddenly became a valuable resource. Waste material handling plants have become big business (see Figure 16-2).

Figure 16-2. Waste material has become a valuable resource.

When governments and major newspapers announced policies of only printing on paper that contained recycled material, trash became valuable. Then in 1994, the United States Supreme Court ruled that private companies must be allowed to compete with local governments in establishing transfer stations. Transfer stations are facilities where small amounts of refuse are dumped, compacted, and shipped to distant landfills in distant areas.

NOTE

When governments and major newspapers announced policies of only printing on paper that contained recycled material, trash became valuable.

Permits are required by all the model fire prevention codes for the operation of waste material and rubbish handling facilities. This includes recycling facilities where newspapers are separated and baled for shipping.

Work closely with your building official to ensure that building use group, sprinkler system design, and building separation issues are addressed. The establishment of these facilities can be particularly political and contentious. Extra attention to the details is well worth the effort.

The illegal storage of waste tires in buildings poses an extra challenge. The obvious code violations are compounded by concerns for firefighter safety in the event of a fire, and in many cases, the criminal nature of the storage. Illegal tire dumping is a felony in some jurisdictions. Close coordination between governmental agencies is necessary. The safety of firefighters and the public must be the primary objective.

NOTE

The illegal storage of waste tires in buildings poses an extra challenge; the obvious code violations are compounded by concerns for fire fighter safety.

LUMBERYARDS AND WOODWORKING PLANTS

Fires in lumberyards are especially challenging for firefighting forces. Lumber is stacked with spacers called "stickers" (see Figure 16-3) to enable forklift operators to quickly remove a large number of boards at once. These spacers create flue spaces in the event of a fire. The sheer volume of the exposed surface area of the lumber piles is conducive for radiant heat transfer between piles. Emergency apparatus access is critical because large quantities of water are needed and are applied with deck guns and elevated streams.

The model fire codes attempt to minimize the hazards associated with lumber storage by:

- Limiting piles to 150,000 cubic feet.
- Limiting lumber piles to 20 feet in height.
- Requiring driveways on 50 foot by 150 foot grids.
- Requiring a 6-foot security fence around the storage.
- Requiring an adequate water supply and fire protection equipment.
- Prohibiting smoking.
- Requiring the removal of combustible waste and vegetation.

NOTE

The sheer volume of the exposed surface area of the lumber piles is conducive for radiant heat transfer between piles.

Figure 16-3. Lumber is stacked to facilitate moving with forklifts. This creates horizontal flue spaces.

Permit requirements for lumberyards are based on the number of **board feet** (fbm) stored at the site, not cubic feet. To determine whether a facility needs a permit while you are in the field, use the following rule of thumb. Since:

    A board foot is 1 foot by 1 foot by 1 inch thick, and

    A cubic foot of wood contains 12 board feet (1 foot by 1 foot by 12 inches), then

    A lumber pile 20 feet by 20 feet by 20 feet in height = 20 × 20 × 240 (from 20 feet × 12 inches) = 96,000 fbm.

Add the piles together. Anything bigger than 20 feet by 20 feet by 20 feet requires a permit. Very few facilities will be within a questionable range. If they are in the lumber business, they will have well over 100,000 board feet; if not, they will generally have considerably less.

Woodworking plants are potentially hazardous occupancies due to the quantity of combustible wastes that are generated. Large volumes of sawdust and shavings are created by woodworking equipment. Automatic waste removal systems must comply with the requirements of the mechanical code and standards referenced by your model fire prevention code. At a minimum, floors must be swept at the end of each shift and floor sweepings must be removed from the building or stored in closed metal bins.

NOTE

Woodworking plants are potentially hazardous occupancies due to the quantity of combustible wastes that are generated.

Inspectors should examine machinery and interior surfaces within the facility for evidence of sawdust accumulation. Sawdust must be removed from interior surfaces including bar joists and rafters. Accumulations may indicate that the waste removal system is not functioning properly. Properly sized and maintained fire extinguishers are especially important in these facilities.

Summary

Large loss fires in industrial and storage facilities have a significant economic impact on society at large. We all pay through increased costs for consumer goods and services, insurance, and consumer credit. Effective fire prevention programs implemented by state and local governments, the insurance industry, and the business community can reduce the risk of fire.

General precautions as well as specific requirements from the fire prevention code must be enforced. Inspectors should verify that a facility has not changed to a more hazardous use and that fixed fire suppression systems are still adequate.

Review Questions

1. List six factors that have been identified as contributing to large losses in industrial occupancies.

    a. _____
    b. _____
    c. _____
    d. _____
    e. _____
    f. _____

2. List five causes or contributing factors to fires in storage occupancies.

    a. _____
    b. _____
    c. _____
    d. _____
    e. _____

3. The original hazard classification of a sprinklered warehouse can be determined by checking the _____ listed on the hydraulic nameplate located on the sprinkler riser.

4. The maximum height of idle pallet storage that can be protected by an NFPA 13 ordinary hazard sprinkler system is _____.

5. In industrial occupancies, smoking should only be permitted _____.

6. List four ways in which the model fire prevention codes address the hazards associated with lumber storage.

    a. _____
    b. _____
    c. _____
    d. _____

7. 1 board foot measures _____ by _____ by _____ thick.

8. Woodworking plants are potentially hazardous occupancies due to _____.
9. At a minimum, floors in woodworking plants should be swept _____.
10. Accumulations of sawdust on bar joists and rafters may indicate that _____.

Discussion Question

1. During an inspection of a fully sprinklered cabinet-making shop of noncombustible/unprotected construction, you find that a small travel trailer has been moved in and is being used as an office. The wheels have been removed, and it has been connected to the building electrical service. The owner sheepishly informs you that he knew you would have a problem with the lack of sprinkler protection in his new office and offers to immediately have it installed.

    a. What code violations exist regarding the trailer?
    b. What would be your actions?

KEY TERMS

board foot

a measurement used for lumber equaling 144 cubic inches (12 inches by 12 inches by 1 inch)

Chapter 17

Gases and Cryogenic Fluids

Learning Objectives

Upon completion of this chapter, you should be able to:

- Describe the four physical states in which gases are commonly stored within containers.
- Define the term gas.
- Describe the Combined Gas Law and its significance.
- List two fire safety strategies for gas storage as they relate to the Combined Gas Law.
- Explain why LPG container storage requirements are expressed in water gallon capacity.

THE PHYSICAL PROPERTIES OF GASES

Solids retain their size and shape. Liquids have a definite size and at least one shape, a flat surface. But gases have neither. No size, no shape and, seemingly, no rhyme or reason.[1]

James H. Meidl

It would be hard to imagine our lives without **gases** and **cryogenic fluids**. Compressed gas refrigeration systems cool our buildings and preserve our foods. Anhydrous ammonia is used to fertilize our croplands. Gases are used as anesthetics during surgical procedures, while oxygen supports our breathing during surgery. Steel fabrication, chemical processing, and even fixed fire protection systems depend on gases. Your soft drink at lunch would have been pretty awful without the bubbles provided by carbon dioxide gas.

Like flammable and combustible liquids, the same characteristics that make gases and cryogenic fluids useful can become deadly attributes with misuse or mishandling. High heat output in combustion, high reactivity in chemical processing, extremely low temperatures, and the economy of handling large volumes easily in compact form are attributes that make compressed gases invaluable in our everyday lives.

The term *gas* refers to a state of matter. For our purposes, a gas is any substance that boils at atmospheric pressure and at any temperature between the absolute zero of outer space and approximately 80°F.[2] Of the ninety-two natural elements, eleven are gases under standard conditions[3]: hydrogen, nitrogen, oxygen, fluorine, chlorine, and the noble gases—helium, neon, argon, krypton, xenon, and radon, which are inert. An apparently unlimited number of compounds, such as air, also have similar boiling points and thus exist as gases.

NOTE

A gas is any substance that boils at atmospheric pressure and at any temperature between the absolute zero of outer space and approximately 80°F.

Some gases and gas mixtures are flammable, others are toxic; some are reactive, some are inert. Some pose multiple hazards. Some are colorless and odorless; some are heavier than air and will settle and form pockets, waiting for an ignition source. Inspectors must consider the relative hazards based on the physical properties of the gas or gas mixture. Good practice and compliance with the code provisions for a nonflammable compressed gas may be totally inadequate for a toxic gas or a gas that is an oxidizer.

> **NOTE**
>
> **Good practice and compliance with the code provisions for a nonflammable compressed gas may be totally inadequate for a toxic gas or a gas that is an oxidizer.**

The Physical Laws of Gases

Somewhere in your educational past, an instructor probably explained a series of scientific laws or facts about gases. The physical properties of gases are the basis for the requirements of the model fire prevention codes, so let us review a moment.

Boyle's law[4] states that the volume occupied by a given mass of gas varies inversely with the absolute pressure if the temperature remains constant.

Charles's law[5] states that the volume of a given mass of gas is directly proportional to the absolute temperature if the pressure remains constant.

The Combined Gas Law,[6] which combines Boyle's and Charles's laws, states that the pressure of a gas is dependent upon the volume of the container it is in and the temperature it is exposed to.

> **NOTE**
>
> **The pressure of a gas depends on the volume of the container it is in and the temperature it is exposed to.**

Gaseous States in Storage

Gases in storage containers exist in one of four states:

- Nonliquefied gases, which are stored under their charged pressure and are entirely gaseous at 68°F.
- Liquefied gases, which are stored under their charged pressure and are partially liquid at 68°F.
- Compressed gases in solution, which are nonliquefied gases dissolved in a solvent.
- Cryogenic fluids, which are liquefied refrigerated gases with boiling points ranging between −130°F and absolute zero (−460°).

Strategies for Safety

With the physical properties of gases in mind, consider the basic strategies that should be used to ensure the safe storage, use, and handling of compressed and liquefied gases:

- Keep the temperature constant so the pressure does not increase.
- Control combustible construction, combustible storage, and vegetation to limit fire exposure.
- Provide a fixed fire protection system to cool containers.
- Limit quantities in storage and use to limit the exposure potential.
- Increase the volume of the container if the temperature increases.
- Provide overpressure protection such as pressure relief valves and fusible plugs.
- Protect against container failure at normal temperatures.
- Require all containers, valves, and fittings to conform to a standard.
- Require periodic hydrostatic testing of containers.
- Require that valves and containers be protected against falling and physical damage (see Figure 17-1).

- Protect against the specific physical hazards of the gas.
- Provide ventilation, fixed fire protection, and emergency containment.

- Develop specific code requirements that address flammability, reactivity, toxicity, and special hazards.

Figure 17-1. This CNG fueling facility has vehicular protection and no combustible material in proximity. Note the cylinders are secured.

CODE REQUIREMENTS FOR GASES

Code requirements applying to gases are found within the model fire prevention codes in various articles and chapters, including Compressed Gases, Cryogenic Liquids or Fluids, Welding and Cutting, Fruit Ripening Processes, Refrigeration, Hazardous Materials, Liquefied Petroleum Gas, Natural Gas Systems, and Fumigation. These articles and chapters have referenced standards that range in scope from bulk oxygen systems at health-care facilities to compressed natural gas as a vehicle fuel.

General requirements for all gases include:

- All containers, cylinders, pressure vessels, valves, and piping must be designed, constructed, and maintained in accordance with national standards including ANSI, DOT, and NFPA.
- All containers must be identified and marked in accordance with DOT regulations.
- Containers must be protected against physical damage and falling with adequate barrier protection or bracing, or by **nesting**.
- Liquefied compressed gas cylinders and containers and cylinders of gas dissolved in solvent must be stored in an upright position.
- Containers must be protected against exposure from fire by maintaining clearance from combustibles and clearing vegetation.
- Adequate separation must be maintained between gas storage and other hazards, including other gases.
- Limits on quantities by building use group and as specified in the fire prevention codes must be maintained.

NOTE

All containers must be identified and marked in accordance with Department of Transportation regulations. Containers must be protected against physical damage and falling with adequate barrier protection or bracing, or by nesting.

CONSIDERATIONS FOR SPECIFIC GASES AND FLUIDS

Liquefied Petroleum Gas

Liquefied petroleum gas (LPG) is not a specific substance, rather, it is a mixture of gases. LPG is composed primarily of propane and butane with small amounts of propylene, ethane, ethylene, and other hydrocarbons[7] and is never the same mixture.

NOTE

Liquefied petroleum gas (LPG) is not a specific substance, rather, it is a mixture of gases.

Throughout the codes, quantity limitations of gases are specified in cubic feet of gas. Because the physical properties of the constituent gases that make up LPG are different, as shown in Table 17-1, the standard system is impossible. One hundred cubic feet of

LPG today and 100 cubic feet of LPG tomorrow will not be the same amount of gas, because you never have the exact same mixture twice. This problem is remedied by using a constant that never changes. LPG quantity limits for tanks and containers are specified by the capacity of the vessel in water gallons.

NOTE

LPG quantity limits for tanks and containers are specified by the capacity of the vessel in water gallons.

Table 17-1 Properties of LPG Constituent Gases

GAS	Butane	Propane	Ethylene	Propylene
Density at ambient temperature (lb/ft^3)	0.15537	0.11599	0.07870	0.11044
Weight per gallon	4.865	4.223	4.735	4.343
Specific volume (ft^3/lb)	6.3356	8.4515	12.7	8.875

Source: Compressed Gas Association, *Handbook of Compressed Gases* (New Source: Compressed Gas Association, Handbook of Compressed Gases (New York: Van Nostrand Reinhold, 1981), page 243.

Because the **vapor densities** of butane and propane are 2.0 and 1.6, respectively, LPG is heavier than air and any leak normally finds its way to the lowest area or part of the building. Prohibitions within the model fire prevention codes and referenced standards against the use of containers and first stage regulating equipment within buildings or on the roofs or balconies of buildings are based in part on the vapor density of the gas mixture.

NOTE

LPG is heavier than air and any leak normally finds its way to the lowest area or part of the building.

The proliferation of LPG barbecue grills in residential occupancies, especially multifamily occupancies, poses a formidable challenge to fire suppression forces as well as fire prevention bureaus. Cylinder exchange programs, in which consumers return empty containers for full containers have created the hazardous situation of consumers carrying partially filled containers into busy mercantile occupancies seeking information on exchanging them (see Figure 17-2). Fire inspectors should require facilities that offer this service to prominently post warnings on the front entrance that LPG cylinders cannot be brought into the building.

Figure 17-2. Consumers bringing LPG cylinders into mercantile occupancies pose a severe hazard. (Courtesy of Duane Perry)

Medical Gases

Medical gases, including flammable anesthetics and oxygen, must be stored in accordance with the general requirements for all gases, as well as the requirements of NFPA 50 and NFPA 99. Fixed piping systems cannot be used to distribute flammable medical gases such as anesthetics. Only nonflammable gases may be distributed through fixed piping systems.

Toxic Gases

Toxic gases are severely limited in use and storage. Exempt quantities are such that few facilities other than Use Group H structures will have much around. Requirements for gas cabinets, exhausted enclosures, treatment systems, and other safety requirements are extensive.

As we discussed in Chapter 12, employees must be trained in the relative hazards, receive training in handling emergencies, and be fully familiar with breathing apparatus and emergency equipment provided. Every inspection should include a review of training records and inspection of the readiness of breathing apparatus and emergency equipment. As you inspect a facility, ask an employee when was the last time he or she actually donned a breathing apparatus.

> **NOTE**
>
> **As you inspect a facility, ask an employee when was the last time he or she actually donned a breathing apparatus.**

Fumigation operations in which toxic gases are used to eliminate insect infestations have specific code requirements that include the posting of the property and the stationing of sufficient, capable alert watchmen, with breathing apparatus, to guard against the entry of any person. Inspectors should check the condition of breathing apparatus and verify that the watchmen have communications capability such as a cell phone, or have access to a telephone. They must understand that they are to remain awake and alert.

Anhydrous Ammonia

Anhydrous ammonia is used in agriculture as a fertilizer. It is a key element in explosives manufacturing. It is also used in building refrigeration systems and in blueprinting. It is flammable, but its flammable range is extremely high: a 16 percent concentration with air is needed to reach the lower flammable limit.[8] It is a corrosive gas, so quantities are limited to the exempt amounts in all use groups, except the appropriate high hazard prescribed by your model building code. Leaks from blueprinting equipment illegally installed within business use buildings pose a significant risk to the occupants. Anhydrous ammonia is particularly hazardous to those with respiratory problems.

Ethylene Gas

Fruit and crop ripening with ethylene gas poses two hazards. Ethylene is flammable and has an anesthetic effect. The heating of buildings used for fruit and crop ripening must be through indirect methods such as steam or hot water, or with sealed listed heating equipment.

Acetylene

Welding and cutting using acetylene is only permitted in areas approved for the purpose, such as factory/industrial occupancies or areas that have been made safe for such operations. Permits required by the model fire prevention codes are to cut and weld in areas other than factory/industrial type areas. The permit requirement enables the fire official to make the provisions of the code clear to the contractor. It also enables the fire official to inform the contractor that if he does not follow the rules, his permit will be revoked and he is out of business in the jurisdiction. This might seem heavy-handed until a contractor burns the roof off of a big-box mercantile occupancy. His violation? Failure to have a water extinguisher or garden hose on hand and failure to maintain a fire watch.

Welding and cutting within occupied buildings is dangerous business. The model codes recognize the relative hazards. The IFC requires the area to be free of combustibles and the floor swept clean for a radius of 35 feet. A capable fire watcher with fire extinguishing capability must be posted and remain for at least 30 minutes after operations are complete. Acetylene cylinders should be stored away from oxygen cylinders, and clearances required by the codes should be maintained. Acetylene cylinders must be maintained in the upright position. An upright position is considered to be within 45 degrees of vertical.

> **NOTE**
>
> **Acetylene cylinders must be maintained in the upright position, which is considered to be within 45 degrees of vertical.**

Cryogenic fluids

Cryogenic fluids are gases that are liquefied by cooling them below their boiling points.

To put the process into perspective, consider this: Gasoline has a boiling point that is somewhere between 100 and 400°F. Its flash point is considerably lower, about −45°F. Methane, the principal component of natural gas (NG) has a boiling point of −289°F. With a boiling point that low, its flash point really is not important. Cool methane below −289°F and it becomes a liquid and can be stored and transported easily in small containers. Cryogenic fluids must be refrigerated to maintain them in the liquid state and prevent pressure buildups.

Cryogenic fluids present hazards in three basic ways. First is the hazard posed by the substance itself. Flammability, toxicity, reactivity, or action as an oxidizer are all possibilities. Second, cryogenic fluids have incredible liquid-to-vapor expansion ratios. The liquid-to-gas ratio of cryogenic nitrogen is 697 to 1![9] Gases that are not toxic can quickly kill by asphyxiation. Third is the hazard posed by the extreme cold of the fluid. Injuries to the human tissues from the extreme cold resemble burn injuries. The tissue is destroyed.

Summary

Matter exists in three states—solid, liquid, and gaseous. Gases assume the shape of whatever container or space they are introduced into. The Combined Gas Law, based on Boyle's and Charles's laws, states that the pressure a gas exerts upon its container is dependent upon the volume of the container and the heat that the gas is exposed to. The model fire codes use storage limitations, protection from exposures, and pressure relief to address the hazards posed by gases described by these laws of physics.

Gases are the only totally capable and ready fuels (those that are flammable). They require no preheating. The only limitations to their flammability are the amount of oxidizer available and the heat required for ignition.

Gases are stored in four basic physical states: nonliquefied, liquefied, compressed gases in solution, and cryogenic fluids.

Review Questions

1. List the four states in which gases are stored in containers.

    a. _____
    b. _____
    c. _____
    d. _____

2. The Combined Gas Law states: _____
3. List four safety strategies used in dealing with compressed and liquefied gases.

    a. _____
    b. _____
    c. _____
    d. _____

4. A method of securing compressed gas cylinders in groups in which each cylinder has a minimum of three points of contact with other cylinders, walls, or bracing is called _____.
5. Liquefied compressed gas cylinders and cylinders of gas dissolved in solvent must be stored _____.
6. Because LPG is a mixture of gases with different physical properties, tank and container size is expressed in terms of _____.
7. Gases with vapor densities greater than 1 are _____ than air.
8. Fire inspectors should require facilities that offer LPG cylinder exchange service to _____.
9. Watchmen at the scene of fumigation operations must have _____ and _____.
10. _____ are gases that are liquefied by cooling them below their boiling points.

Discussion Questions

1. Why are the threads on flammable compressed gas cylinders designed to require attachment using a counterclockwise motion?
2. You have arrived on the site of a house to be fumigated with a toxic gas to eliminate an insect infestation. The contractor has applied for a permit as required by the fire prevention code. He advises that his employee who will act as the watchman have access to a pay phone on the next block and a self-contained breathing apparatus (SCBA). Upon questioning the job foreman, you discover that the watchman plans to spend the night in a small tent in the front yard of the residence. When you ask the watchman to allow you to examine the SCBA, you realize that he does not speak English.

a. Does this operation meet the requirements of your fire prevention code?
b. Should the permit be issued?

KEY TERMS

Compressed gas

a material or mixture of materials which:

- is a gas at 68°F or less at 14.7 psia of pressure
- has a boiling point of 68°F or less at 14.7 psia, which is either liquefied, nonliquefied, or in solution; those gases that have no other health- or physical-hazard properties are not considered to be compressed until the pressure in the packaging exceeds 41 psia at 68°F

cryogenic fluid

a refrigerated liquid gas with a boiling point below −130°F

nesting

method of securing compressed gas cylinders in groups, in which each cylinder has a minimum of three points of contact with other cylinders, walls, or bracing

vapor density

the ratio of the weight of a given volume of a gas to that of air

Chapter 18
Pesticides and Other Health Hazards

Learning Objectives

Upon completion of this chapter, you should be able to:

- Identify the signal words required for all Class I and Class II pesticides.
- List the three categories of pesticides.
- Describe the multiple hazards posed by many pesticides.
- Describe the code provisions that ensure employee and firefighter awareness of the presence of toxic materials.
- Describe the difference between a toxic material and a health hazard material.

PESTICIDES

The insect and rodent damage to stored grain and foods each year exceeds $500 million in the United States alone.[1] Pesticides are a part of our everyday lives. We all experience low-dose exposures to pesticides on an almost daily basis, but most reported injuries are to farmers, loggers, and migrant farm workers.[2] Pesticides are in almost every household and are sold in nearly every grocery, drug store, and hardware store. Class I and Class II pesticides are covered by specific pesticide storage and display provisions in the model fire prevention codes. Class I and II pesticides are identified by the signal words "POISON," "DANGER—POISON," or "DANGER" printed on the containers in accordance with the Federal Insecticide, Fungicide and Rodenticide Act (FIFRA). If the signal words are not on the container, the material is not covered by the pesticide provisions.

> **NOTE**
>
> **Class I and II pesticides are identified by the signal words "POISON," "DANGER–POISON," or "DANGER."**

Pesticides are divided into three groups: rodenticides, herbicides, and insecticides. All are designed to kill and can kill or injure people, pets, and wildlife as effectively as they kill harmful insects, rodents, and plant life. Accidental poisoning can occur by absorption through the skin and tissues, ingestion, or inhalation. When exposed to fire, pesticides can be carried by runoff from hose lines and sprinkler systems and carried with smoke and fire gases, exposing persons both inside and outside the structure.

> **NOTE**
>
> **Accidental poisoning can occur by absorption through the skin and tissues, ingestion, or inhalation.**

Many pesticides pose multiple health and physical hazards, and are covered by multiple chapters or articles of the fire codes. Pesticides are often also classified as flammable or combustible liquids, oxidizers and toxic and highly toxic liquids, solids, and compressed gases. In addition to the fire prevention code requirements that pertain to all flammable and combustible liquids, oxidizers and all toxic and highly toxic solids, liquids, and compressed gases, pesticide storage and handling must comply with the following:

- All containers and packages must be marked.
- Pesticide storage is limited to the first floor, with direct access to the outside.
- Storage rooms must be marked in accordance with the NFPA 704 system.
- Incompatible materials must be effectively segregated.
- Pesticides cannot be stored in the same area with ammonium nitrate fertilizer.
- Storage areas must be designed to contain hose stream runoff.

- Storage and display shelving must be stable and, where practical, secured to the wall.

- Retail displays must conform to maximum square foot densities, maximum container sizes, and maximum display heights.

NOTE

Many pesticides pose multiple health and physical hazards, and are covered by multiple chapters or articles of the fire codes.

When calculating quantities to determine conformance with exempt quantity requirements for use groups or the need for permits, remember that these materials may pose multiple hazards. A quantity of a pesticide that is a highly toxic liquid and a combustible liquid will be counted twice and added to the inventory under each category.

TOXIC AND HIGHLY TOXIC MATERIALS

The code definitions for toxic and highly toxic solids, liquids, and gases are derived from federal hazardous materials regulations promulgated by the Department of Labor and are expressed in terms of LD50 and LC50. These refer to doses (LD50) by skin absorption or oral ingestion that proved fatal within 1 hour to 50 percent of the laboratory animals, or concentrations (LC50) in air that had the same results.

Notification, Warning, and Employee Training

All areas where hazardous materials are stored, handled, or used have specific requirements for identification with NFPA 704. MSDS sheets for all hazardous materials must be on site regardless of quantity. Repository containers for fire department use are required over certain thresholds. The requirement for MSDS sheets at buildings below these thresholds is for employee and public safety in the event of a spill or accidental poisoning.

NOTE

MSDS sheets for all hazardous materials must be on site regardless of quantity.

Employees who work with hazardous materials are required to be aware of the potential dangers and be trained in appropriate response to accidents or emergencies. Every inspection should include verification that employee training programs are being conducted as required. Like all records required by the model fire codes, if you wait for the accident or fire to ask for them, they may not be there!

Summary

Toxic and highly toxic solids, liquids, and gases— commercial pesticides fall into one of these categories. Most pesticides are also a flammable or combustible liquid, compressed gas, or an oxidizer. Pesticides pose multiple hazards. Each hazard must be addressed through the applicable provisions of the code.

Review Questions

1. Class I and II pesticide can be identified by the signal words: _____, _____, or _____.
2. If the signal words are not on the label of a pesticide, it is regulated by the model fire code as a _____.
3. List the three types of pesticides.

    a. _____
    b. _____
    c. _____

4. Accidental poisoning can occur through _____, _____, or _____.
5. List four storage and handling requirements for Class I and Class II pesticides.

    a. _____
    b. _____
    c. _____
    d. _____

6. List three of the multiple hazards that pesticides may pose.

    a. _____
    b. _____
    c. _____

7. Employees who work with hazardous materials are required to be trained _____.

8. A chemical that is not corrosive and has a reversible inflammatory effect on living tissue is called a(n) _____.
9. LD_{50} refers to lethal doses by _____.
10. LC_{50} refers to lethal doses by _____.

Discussion Question

1. During an inspection at a large garden supply center, the manager questions the need to provide training for his employees in the appropriate response to spills, releases, or accidents.

    a. How can you verify that a training program is in place?
    b. Are all employees required to be trained?

KEY TERMS

irritant

a noncorrosive chemical that has a reversible inflammatory effect on living tissue at the point of skin contact

David Diamantes

Chapter 19

Using Referenced Standards

Learning Objectives

Upon completion of this chapter, you should be able to:

- Describe what is meant by permissive code language.
- List three features of the standards referenced by the model fire prevention codes.
- List five organizations that develop standards referenced or recognized by the model fire codes.

REFERENCED STANDARDS

In Chapter 1 we discussed code development and adoption. Throughout this text, we have referred to various standards referenced by the model codes. These standards regulate construction, material characteristics and performance, and system installation and performance. Without them, effective enforcement of the model codes is impossible.

Almost 200 different standards are referenced by the 2012 IFC. They range from standards that regulate the maintenance of water-based extinguishing systems (NFPA 25), to above-ground steel tanks for flammable and combustible liquids (UL 142). Each of the model codes contains references to specific standards within the code text. These references cite specific applications for the standards. Within the scope of the cite, a referenced standard is considered part of the code. Advisory provisions and those considered **permissive**, or discretionary for the code official, are not considered **mandatory**.

> **NOTE**
>
> **Within the scope of the cite, a referenced standard is considered part of the code.**

Not all guidelines for construction, materials, and equipment can be adopted by the model codes as reference standards. Only those developed through nationally recognized consensus processes are adopted by the model codes. Standards must be performance based and not specify particular products or materials. Provisions must be mandatory, not permissive, and the standard must be readily available.

Specific editions are referenced by the model codes and listed by year. Code officials may only enforce the provisions of the edition referenced by the model code, or as amended by the adopting ordinance.

Purchasing Referenced Standards

Code books and referenced standards are expensive-- consider them the tools of the trade. Imagine a surveyor without a transit! Without access to the complete set of the standards referenced by your model fire code you do not have the tools to do the job. Every inspector does not need to carry all one hundred or so documents in the back seat of his car. They should, however, carry a fire and building code in the field. Depending on the jurisdiction, copies of frequently used standards should also be carried in the field. Standards for sprinkler system installation, water-based extinguishing system maintenance, and flammable and combustible liquid storage and handling are a few that are frequently needed.

A complete set of all of the referenced documents should be available at the office, for the use of code enforcement personnel as well as citizens who may have questions. The public has a right to view the code documents. They may not understand them, but they have a right to read the specific provisions that apply to their situation.

Code books and referenced standards should never be thrown away. Previous editions should be maintained in order to determine continued compliance with previously approved conditions. Subscription services, which provide continual updates to referenced standards and recommended practices, are a convenient method of purchasing these documents. When updates and replacement pages arrive, do not throw away the old pages. You will end up with a newer edition than is referenced in your model code. Although the newest edition can be used for guidance, you cannot enforce it. The fire official may accept compliance with the newer standard as meeting the intent and spirit of the code, if a building owner offers or requests to use the standard. Only the provisions of the referenced edition are mandatory.

Copies of the referenced standards are available from the model code organizations and from the standards organizations themselves. Addresses for the organizations are listed within the model codes.

Summary

Like the model building and fire prevention codes that reference them, national standards are developed through a formal consensus process. The standards are referenced within specific code sections and are listed together within a separate chapter or article. Specific editions of the standards are referenced and thus become an enforceable part of the code. Like nationally developed recommended practices, new editions of national standards can be used for guidance but cannot be enforced.

Review Questions

1. Within the scope of the cite, a referenced standard is considered to be _____.
2. Advisory provisions of the standards are considered to be _____.
3. Provisions that are not discretionary are considered to be _____.
4. Only standards developed through _____ are referenced by the model codes.
5. Specific edition of standards are adopted by reference and listed by _____.
6. Previous editions of codes and standards should be _____.
7. Newer editions of standards than those that are referenced by the model codes may be used by fire officials for _____.
8. May referenced standards specify particular products or materials? _____

9. Using your model building or fire code, list five organizations that develop standards referenced by the codes.

   a. _____
   b. _____
   c. _____
   d. _____
   e. _____

10. Must standards be readily available in order to be adopted by the model codes? _____

Discussion Question

1. NFPA 58, *Standard for the Storage and Handling of Liquefied Petroleum Gases* is adopted by reference by the IFC and the UFC.

    a. If there is a conflict between the model code and referenced standard, which provision should be applied?
    b. If a condition is not addressed within the model code or within the referenced standard, what are the fire official's sources for guidance?

KEY TERMS

permissive

code language that is discretionary and leaves application to the judgment of the enforcing authority

mandatory

code provisions that must be enforced and complied with

Appendix A
Equivalents, Conversion Charts, And Related Mathematics

Volume and Capacity Equivalents

U.S. Gallons	Cubic Meters	Liters	Cubic Inches
1	0.003785	3.785	231.0
1.201	0.004545	4.5	277.41
0.004329	0.00001639	0.01639	1.0
7.481	0.02832	28.32	17.28
264.2	1.0	1000.0	61023.0
0.2642	0.001000	1.0	61.023

Cubic Measure

1728 cubic inches	1 cubic foot
27 cubic feet	1 cubic yard
128 cubic feet	1 cord (firewood)
40 cubic feet	1 ton (shipping)
2150.42 cubic inches	1 standard bushel
231 cubic inches	1 gallon

U.S. Standard Gauge

Gauge	Thickness	Decimal
0	5/16	0.3125
3	1/4	0.2500
7	3/16	0.1875
8	11/64	0.1718
10	9/64	0.1406
12	7/64	0.1093
14	5/64	0.078

Conversion Tables

Multiply	By	To Obtain
Diam. circle	3.1416	Circumference
Diam. sphere cubed	0.5236	Volume
Gallons	0.1337	Cubic feet
Gallons	8.33	Pounds of water
Cubic feet water	62.427	Pounds of water
Water column feet	0.4336	PSI at base
Inches of mercury	0.4912	PSI

Metric Equivalents

1 centimeter	0.3937 inches
1 inch	2.54 centimeters
1 decimeter	3.937 inches
1 decimeter	0.3282 feet
1 foot	3.048 decimeters
1 meter	39.37 inches
1 meter	1.0936 yards
1 yard	0.9144 meters
1 kilometer	0.62137 miles
1 mile	1.6093 kilometers

Measure of Volume

1 cubic centimeter	0.067 cubic inches
1 cubic inch	16.39 cubic centimeters
1 cubic decimeter	0.0353 cubic feet
1 cubic foot	28.317 cubic decimeters
1 cubic meter	1.308 cubic yards
1 cubic yard	0.7646 cubic meters
1 liquid ounce	0.029 liters
1 liter	33.814 liquid ounces
1 liter	1.0567 liquid quarts
1 liquid quart	0.9463 liters
1 gallon	3.785 liters

Miscellaneous Information

To find the diameter of a circle multiply the circumference by 0.31831.

To find the circumference of a circle, multiply the diameter by 3.1416.

The find the area of a circle multiply the square of the diameter by 0.7854.

To find the surface of a sphere multiply the square of the diameter by 3.1416.

Doubling the diameter of a pipe or hose increases its capacity four times.

TRIANGLE

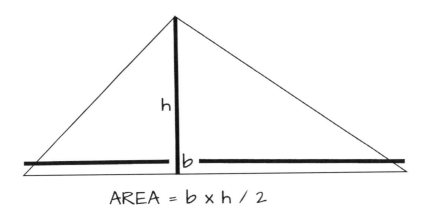

AREA = b x h / 2

RIGHT TRIANGLE

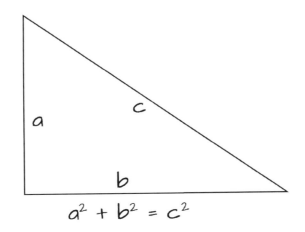

$a^2 + b^2 = c^2$

LENGTH OF SIDES

$a = \sqrt{c^2 - b^2}$

$b = \sqrt{c^2 - a^2}$

$c = \sqrt{a^2 - b^2}$

CIRCLE

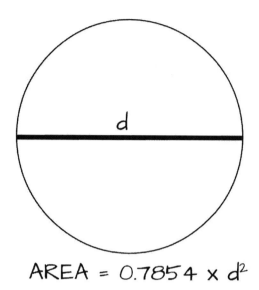

AREA = 0.7854 × d^2

CYLINDER

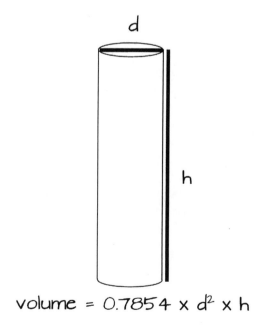

volume = 0.7854 × d^2 × h

RECTANGLE

AREA = a x b

RECTANGULAR BOX

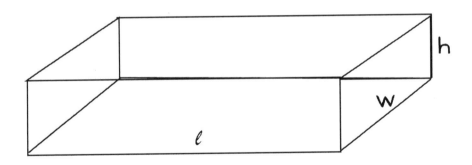

VOLUME = ℓ x w x h

PARALLELOGRAM

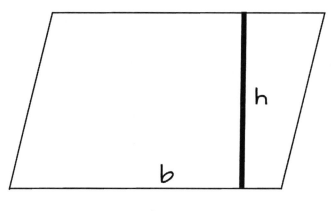

AREA = b x h

NOTES

Preface

1. Jennifer Hemmingsen, The Albert Lea Tribune (Albert Lea, MN), July 9, 2001.
2. *Fire in the United States*, (Emmitsburg, MD: United States Fire Administration, National Fire Data Center, August 1999), p. 1.
3. *America at Risk*, (Emmitsburg, MD: *Recommissioned Panel for America Burning*, Federal Emergency Management Agency, May 2000), p. 15.
4. Ibid., p. 16.
5. National Association of Credit Men, Committee on Fire Insurance, June 1909, quoted in Harry Chase Brearley, *Fifty Years of Civilizing Force* (New York: Frederick A. Stokes, 1916), p. 111.
6. *Twenty-five Fatality Fire at Chicken Processing Plant, Hamlet, North Carolina, September 1991*, (Technical Report #057) (Emmitsburg, MD: United States Fire Administration, June 1999), p. 9.

Fire and Emergency Services Higher Education (FESHE)

1. *2001 Fire & Emergency Services Higher Education Conference Final Report*, (Emmitsburg, Maryland: U.S. Fire Administration, 2001), p. 12.
2. National Fire Academy FESHE Model Curriculum (Emmitsburg, Maryland: U.S. Fire Administration, 2011), p. 12.

Chapter One

1. *Field Practice—An Inspection Manual for Property Owners, Fire Departments and Inspection Offices* (Quincy, MA: National Fire Protection Association, 1922), p. 11.
2. Joseph K. Freitag, *Fire Prevention and Protection as Applied to Building Construction* (New York: Wiley & Sons, 1912), p. 7.
3. Michael P. Dineen, et al., *Great Fires of America* (Waukesha, WI: Country Beautiful Corporation, 1973), p. 28.
4. Harry C. Brearley, *Fifty Years of Civilizing Force* (New York, NY: Frederick A. Stokes, 1916), p. 42.
5. National Commission on Fire Prevention and Control, *America Burning* (Washington, DC: Government Printing Office, 1973), p. 79.
6. *BOCA Basic/National Building Code* (Country Club Hills, IL: Building Officials and Code Administrators International, 1984), p. iv.
7. Kevin Cassidy, *Fire Safety and Loss Prevention* (Stoneham, MA: Butterworth-Heinemann, 1992), p. 56.

8. BOCA: "The International Code Council," *Building Official and Code Administrator Magazine*, March/April 1995, p. 12.
9. National Commission on Fire Prevention and Control, *America Burning*, p. 79.
10. Albert Harkness, "Building Codes: A Historical Perspective," *Building Officials and Code Administrators Magazine*, 26 (March/April 1996), p. 15.
11. Ibid., p. 79.
12. Ibid., p. 79.
13. National Fire Protection Association, *NFPA 100 Years, a Fire Protection Overview* (Quincy, MA: National Fire Protection Association, 1996), p. 9.
14. Ibid., p. 15.
15. *Code for Safety to Life from Fire in Buildings and Structures* (Quincy, MA: National Fire Protection Association, 1994), p. 17.
16. NFPA 251–95, *Standard Methods of Tests of Fire Endurance of Building Construction and Materials* (Quincy, MA: National Fire Protection Association, 1995), p. 2.
17. *NFPA Standards-Making System* (Quincy, MA: National Fire Protection Association, 1996), p. 3.
18. Timothy Callahan, *Fire Service and the Law* (Quincy, MA: National Fire Protection Association, 1987), p. 134.
19. Clay L. Wirt, *Virginia Town and City*, August 1989, p. 12.
20. George Dean et al., *Legal Aspects of Code Administration* (Country Club Hills, IL: Building Officials and Code Administrators, International Congress of Building Officials, Southern Building Code Congress International, 1984), p. 16.
21. Callahan, p. 134.
22. National Fire Academy, *Fire Prevention Organization and Management*, pp. 6–13.

Chapter Two

1. "Fire Hazards," Report of the Virginia Advisory Legislative Council to the Governor and the General Assembly of Virginia, Senate Document 11 (Richmond: Commonwealth of Virginia Division of Purchase and Printing, 1948), p. 5.
2. John R. Hall et al., *Fire Code Inspections and Fire Prevention: What Methods Lead to Success?* (Boston, MA: National Fire Protection Association, 1978) p. viii.
3. George Dean, et al., Legal Aspects of Code Administration (Country Club Hills, IL: Building Officials and Code Administrators, International Congress of Building Officials, Southern Building Code Congress International, 1984), p. 135.
4. Timothy Callahan, *Fire Service and the Law* (Quincy, MA: National Fire Protection Association, 1987), p. 174.
5. Dean et al., p. 85.
6. Callahan, p. 178.
7. *International Fire Code Commentary* (Falls Church, VA: International Code Council, 2002), p. 25.

8. National Fire Academy, Political and Legal Foundations of Fire Protection (Emmitsburg, MD: National Fire Academy, 1992), p. 9-2.
9. Howard Markman et al., Political and Legal Foundations of Fire Prevention, p. 9-3.
10. Ibid., p. 9-3.
11. Callahan, p. 167.
12. Dean et al., p. 80.
13. Ibid., p. 80.
14. George E. Rush, *Dictionary of Criminal Justice* (Guilford, CT: Dushkin, 1986), p. 221.
15. Dean, p. 142.
16. Rush, p. 104.
17. Ibid., p. 104.

Chapter Three

1. *Fatal Fires in Residential Buildings* (2009-2011) (Emmitsburg, MD: United States Fire Administration, National Fire Data Center, May 2013), p. 1.

Chapter Four

1. Ronny J. Coleman et al., *Managing Fire Services* (Washington, DC: International City/County Management Association, 1988), p. 7.
2. Francis L. Brannigan, *Building Construction for the Fire Service* (Boston, MA: National Fire Protection Association, 1971), p. 8.
3. *International Building Code* (Country Club Hills, IL: International Code Council, 2003), p. 85.
4. Michael Martz and Jeremy Redmon, The Richmond Times Dispatch (Richmond, VA, March 28, 2004).
5. Robert C. Thompson, R. A., A.I.A., "New York State: Building a Case for Standards," Standards Engineering Society, 2002.

Chapter Five

1. Fire Inspection Principles Course Guide (Emmitsburg, MD: National Fire Academy, 1993).
2. *Fire Protection Handbook*, 16th ed. (Quincy, MA: National Fire Protection Association), p. 6-76.
3. Ibid., p. 20.

Chapter Six

1. *Fire Protection Handbook*, 16th ed. (Quincy, MA: National Fire Protection Association), p. 17-2.

2. *Fire Protection Handbook*, 13th ed. (Quincy, MA: National Fire Protection Association), p. 16-2.
3. *Fire Protection Handbook*, 16th ed., p. 4–42.
4. Richard E. Hughey, P. E., "Property Protection or Life Safety—Can We Have Both?" Fire Marshal Quarterly, September 1996, p. 20.
5. IFSTA 210, *Private Fire Protection and Detection* (Stillwater, OK: International Fire Service Training Association, 1987), p. 39.
6. Fire Protection Handbook, 13th ed., p. 16–40.
7. NFPA 13, Standard for the Installation of Sprinkler Systems, 2002 ed. (Quincy, MA: National Fire Protection Association), p. 17.
8. Tom Multer and Richard Thieken, "Careful Planning is Necessary in ESFR Sprinkler Installations," Engineering and HVAC Design.
9. Jesse Roman. "Protecting Saint Patrick's," NFPA Journal, September/October, 2014, p. 48.
10. *High-rise Office Building Fire, One Meridian Plaza, Philadelphia, Pennsylvania* (Emmitsburg, MD: United States Fire Administration), November, 1989, p. 15.
11. Edward J. Kaminski, P. E., "Practical Aspects of Wet and Dry Chemical Extinguishing System Inspection and Acceptance," Building Officials and Code Administrators Magazine, 4 (July/August 1995), p. 34.
12. R. T. Leicht, "Commercial Cooking Protection: The Myths and Facts Concerning the UL-300 Standard," Fire Marshals Quarterly, April 1996.
13. *Fire Protection Handbook*, 13th ed., p. 15–23.
14. NFPA 27, *National Fire Alarm Code*, 1996 ed. (Quincy, MA: National Fire Protection Association, 1996), p. 83.

Chapter Seven

1. Fire Journal (Quincy, MA: National Fire Protection Association, August 1977), p. 50.
2. Ibid., p. 50.
3. *Independent Review of the Cook County Administration Building Fire, Executive Summary* (Washington, D.C.: James Lee Witt Associates, September 30, 2004), p. 34.
4. John R. Paradise, "Insecure: Dangers of Immobile Security Bars," NFPA Journal, July/August 2003.
5. *Security Bars and Fire Safety (Quincy*, MA National Fire Protection Association), 1996.
6. Josh McGhee, "Four Children Dead in Roseland Fire, Barred Doors Trapped Family: Neighbors." DNAinfo.com, http://www.dnainfo.com/chicago/20140908/roseland/four-children-killed-roseland-fire, (accessed 11/07/2014)

Chapter Eight

1. "Flammable Decorations, Lack of Exits Create Tragedy at Coconut Grove," Fire Engineering, August, 1977, p. 67.
2. Ibid., p. 72.
3. *Fire Inspection and Code Enforcement* (Stillwater, OK: Fire Protection Publications, 1987), p. 75.
4. David P. Demars, *Fire Investigation Report, Hotel Fire, Las Vegas, Nevada, February 10, 1981* (Quincy, MA: National Fire Protection Association, 1981), p. 17.

Chapter Nine

1. NFPA 550, *Fire Safety Concepts Tree* (Quincy, MA: National Fire Protection Association, 2002), p. 14.
2. Dibble v. Brunty, United States District Court for the Eastern District of Virginia, Civil Action No. 95-356-A.
3. Fire Prevention Code Commentary (Country Club Hills, IL: Building Officials and Code Administrators, 1993), p. 3–5.
4. Alisa Wolf, "Fraternity Fire Kills Five," NFPA Journal, September/October 1996, p. 61.
5. Ibid., p. 61.
6. Kimberly D. Rohr, Christmas Tree Fires (Quincy, MA: NFPA, 2005), p. 1.
7. Abandoned Cold-Storage Warehouse Multi-Firefighter Fatality Fire Worcester, Massachusetts, December 3, 1999 (Emmitsburg, MD: United States Fire Administration, 2005), p. 2.

Chapter Ten

1. J. K. Richardson, "The Reliability of Automatic Sprinkler Systems," Canadian Building Digest, Volume 238, July 1985.
2. Kimberly D. Rohr, "U.S. Experience with Sprinklers," NFPA 2001, p. 48.
3. Ibid., p. 50.

Chapter Twelve

1. NFPA 704, Identification of the Fire Hazards of Materials (Quincy, MA: National Fire Protection Association, 2001).

Chapter Thirteen

1. Compton's Reference Collection, Compton's New Media, Inc., 1995.
2. Fires in or at Service Stations and Motor Vehicle Repair and Paint Shops (Quincy, MA: National Fire Protection Association, 2002).

3. John D. DeHaan, Kirk's Fire Investigation (Englewood Cliffs, NJ: Prentice-Hall, 1991), p. 44.
4. Ibid., p. 36.
5. Richard Best, "$100 Million Fire in K Mart Distribution Center," Fire Journal, March 1983, p. 36.
6. NFPA 30B, Code for the Manufacture and Storage of Aerosol Products (Quincy, MA: National Fire Protection Association, 2003), p. 24.

Chapter Fourteen

1. James H. Meidl, Explosive and Toxic Hazardous Materials (Beverly Hills, CA: Glencoe Press, 1972), p. 51.
2. Robert C. Morehard, Explosives and Rock Blasting (Dallas, TX: Atlas Powder Company, 1987), p. 3.
3. Compton's Interactive Encyclopedia (Compton's New Media, Inc., 1995).
4. Bureau of Alcohol Tobacco and Firearms, Public Affairs Division "*Explosives in the United States,*" May, 1014. (https://www.atf.gov/publications/factsheets/factsheet-explosives-in-the-united-states.html, accessed 02.08.2015)
5. Timothy D. Stark, "*Is Construction Blasting Still Abnormally Dangerous?*"Journal of Legal Affairs and Dispute Resolution in Engineering and Construction, November 2010.
6. Morehard, p. 330.
7. NFPA 495, *Explosive Materials Code* (Quincy, MA: National Fire Protection Association, 2001), p. 23.
8. Morehard, p. 5.
9. Meidl, p. 4.
10. Morehard, p. 300.
11. T. S. Bajpayee et al., "*An Analysis of Flyrock Accidents in Surface Blasting Operations,*" Proceedings of the 30th Annual Conference Explosives and Blasting Technology (New Orleans, LA: February 1–4, 2004), p. 4.
12. Alert Bulletin, Multiple-death Fire Fighter Fatality Incident Kansas City, Missouri, National Fire Protection Association, December 7, 1988.
13. Ibid.

Chapter Fifteen

1. Alisa Wolf, "Seventeen Die in Dusseldorf Airport Terminal Fire," NFPA Journal, July/August 1996.
2. International Fire Code Commentary (Falls Church, VA: International Code Council, 2002), p. 393.

Chapter Sixteen

1. Stephen G. Badger, "2011 Large-Loss Fires in the United States," (Quincy, MA: National Fire Protection Association, 2012), p.58.
2. Before the Fire—Fire Prevention Strategies for Storage Occupancies (Quincy, MA: National Fire Protection Association, 1988), p. 8.
3. Jack Yates, "Chicken Processing Plant Fires," U.S. Fire Administration, Report #57, p. 11.
4. Ibid.
5. James L. Linville et al., Industrial Fire Hazards Handbook (Quincy, MA: National Fire Protection Association, 1984), p. 974.
6. Ibid., p. 986

Chapter Seventeen

1. James H. Meidl, Flammable Hazardous Materials (Beverly Hills, CA: Glencoe Press, 1970), p. 93.
2. Compressed Gas Association, Handbook of Compressed Gases (New York: Van Nostrand Reinhold, 1981), p. 5.
3. John D. DeHaan, Kirk's Fire Investigation (Englewood Cliffs, NJ: Prentice-Hall, 1991), p. 371.
4. NFPA Fire Protection Handbook, 18th ed. (Quincy, MA: National Fire Protection Association), pp. 4–72.
5. Ibid.
6. Meidl, p. 96.
7. Ibid., p. 45.
8. Ibid., p. 134.
9. Ibid., p. 146.

Chapter Eighteen

1. George W. Rambo, "Pest Control," New York Public Library Desk Reference, Compton's New World, 1996.
2. Richard Stilp and Armando Bevelacqua, Emergency Medical Response to Hazardous Materials Incidents (Albany: Delmar Publishers, 1997), p. 274.

David Diamantes

Glossary

Accelerator A quick opening device that permits system air pressure to enter the dry pipe valve below the one-way clapper valve, unbalancing the differential and causing the valve to trip more quickly.

Accessible means of egress A means of egress including the exit access, exit, and exit discharge that can be entered and used by a person with a severe disability using a wheelchair and is also safe and usable for people with other disabilities.

Accessory use area A portion of a building with a different use group classification than the main area, but that does not require a fire separation between the main area, due to its small size.

Adopt To formally accept and put into effect.

Adoption by reference Method of code adoption in which the specific edition of a model code is referred to within the adopting ordinance.

Adoption by transcription Method of code adoption in which the entire text of the code is published within the adopting ordinance.

Aerosol A product that is dispensed from a container by means of a liquefied or compressed gas.

Ambulatory Able to walk about without assistance. Capable of sensing an emergency situation and appropriately responding by exiting the building.

Approved Acceptable to the code official with jurisdiction. Approval is normally based on nationally recognized standards or, in their absence, on sound engineering practice.

Assembly (rated assembly) A building component such as a door, wall, damper, or ceiling composed of specific parts and tested and listed as a unit.

AST Aboveground storage tank for regulated liquids.

Blasting agents Materials or mixtures containing fuel and an oxidizer not otherwise classified as an explosive and that cannot be detonated by means of a #8 blasting cap when unconfined.

Board foot A measurement used for lumber equaling 144 cubic inches (12 inches by 12 inches by 1 inch).

Building height According to the IBC, "the vertical distance from grade plane to the average height of the highest roof surface."

Citation A written order, issued by a law enforcement officer or other authorized official, directing an alleged offender to appear in court at a specific time to answer a criminal charge.

Closure Tank closure Placing a tank permanently out of service by removal or abandonment in place using an approved method.

Code A systematically arranged body of rules. When and where to do or not to do something.

Combustible liquid A liquid having a flash point at or above 100°F. Combustible liquids are further categorized based on flash point as types II, IIIA, and IIIB.

Common path of travel The portion of an exit access that building occupants must traverse before two distinct paths of travel or two exits are available.

Compressed gas A material or mixture of materials which is a gas at 68°F or less at 14.7 psia of pressure; and has a boiling point of 68°F or less at 14.7 psia, which is either liquefied, nonliquefied, or in solution; those gases that have no other health- or physical-hazard properties are not considered to be compressed until the pressure in the packaging exceeds 41 psia at 68°F.

Conflagration A large fire involving multiple buildings that defies control efforts and causes extensive damage over a large area.

Container A vessel with a capacity of 119 gallons or less used for the storage or transportation of flammable or combustible liquids. Piping and engine fuel tanks are not containers.

Control areas Areas within a building in which hazardous materials in quantities not exceeding the exempt amounts may be stored, handled, or used.

Cryogenic fluid A refrigerated liquid gas with a boiling point below −130°F. Also known as cryogenic liquid.

Dead end A corridor, hallway, or passageway open to a corridor that can be entered from the exit access without passage through a door, but that does not lead to an exit.

Deflagration Very rapid but subsonic oxidation evolving heat, light, and a low-energy pressure wave that is capable of causing damage.

Deluge sprinkler system An automatic sprinkler system that features open heads, dry piping, and a deluge valve that controls the supply of water, designed to wet the entire area upon activation.

Deluge valve Sprinkler water supply valve that is activated automatically or manually.

Density Refers to sprinkler density calculated by gallons per minute discharge divided by the square footage covered.

Detection device A device connected to a fire alarm system having a sensor that responds to physical stimulus such as heat or smoke.

Detonator A device, consisting of electric and nonelectric blasting caps, fuse caps, and detonating cord delay connectors, that contains a primary or initiating explosive designed to set off an explosive reaction.

Detonator, delay type A detonator that introduces a specific time lapse between the firing signal and detonation of the main explosive charge.

Detonator, instantaneous type A detonator with no time lapse between the firing signal and the detonation of the main explosive charge.

Differential Ratio of air pressure to water pressure that is necessary to balance a dry pipe valve, maintaining it in the closed position.

Dillon's rule Legal ruling issued by Chief Justice John Forrest Dillon of the Iowa Supreme Court in the late 1800s whereby local governments possess only those powers expressly granted by charter or statute.

Dry pipe sprinkler system An automatic sprinkler system that features dry piping maintained under constant air pressure and a dry pipe valve in which water is held back by the pressure differential between the system air pressure and water supply pressure. Used where there is a danger of freezing.

Dry pipe valve Sprinkler water supply valve designed to permit a moderate amount of air pressure above the valve to hold back a much greater water pressure from the incoming supply.

Early suppression fast response sprinklers (ESFR) A type of fast response sprinkler designed to suppress fires in high challenge fire hazards through the application of increased flow densities.

Exhauster A quick opening device used in dry pipe sprinkler systems that uses an auxiliary valve to discharge system air pressure to the atmosphere.

Exit That portion of a means of egress that is separated from all other parts of a building by rated assemblies and provides a protected path to the exit discharge.

Exit access That portion of a means of egress from any point in a building to an exit.

Exit discharge That portion of a means of egress between the exit and public way.

Explosive materials Explosives, blasting agents, and detonators, including dynamites, slurries, emulsions and water gels, black powder, smokeless powder, detonators and safety fuses, squibs, detonating cord, and other materials whose primary function is to function by explosion.

Explosives Chemical compounds or mixtures whose primary function is to function by explosion and that cause a sudden and almost instantaneous release of pressure, gas, and heat.

Fire Rapid oxidation accompanied by the evolution of heat and light.

Fire limits or fire districts Designated areas within a jurisdiction where combustible construction is prohibited or restricted in order to reduce conflagration potential.

Fire line Dedicated underground supply piping for a sprinkler or standpipe system.

Fire protection system Equipment and devices designed to detect a fire, sound an alarm and (or) make notification, control or remove smoke and hot gases, and control or extinguish the fire.

Fire resistance The resistance of a building to collapse or to total involvement in fire; the property of materials and their assemblies that prevents or retards the passage of excessive heat, hot gases, or flames under conditions of use.

Flammable liquid A liquid having a flash point below 100°F and further categorized based on flash point and boiling point as types IA, IB, and IC.

Flashover Point at which the contents of a room or space becomes heated, simultaneously ignites, and the entire room or space becomes involved in fire.

Flash point The minimum temperature at which a liquid gives off sufficient vapor to form an ignitable mixture at the surface, but not sufficient to sustain combustion.

Fly-rock Rock propelled from the blast area by blasting operations.

Freedom of Information Act A law passed by Congress and signed by President Lyndon Johnson in 1967. The law, and state laws patterned after it, guarantee public access to all documents and information under control of the government, with certain specific exemptions.

Grade plane An imaginary plane representing the finished ground around the exterior walls of a building.

Halogenated extinguishing agent (halon) A clean extinguishing agent composed of carbon and one or more elements from the halogen series (fluorine, chlorine, bromine, and iodine), which leaves no residue.

Hazardous Materials Inventory Statement (HMIS) An inventory of regulated materials in a form approved by the fire official that includes the following information: chemical name; manufacturer's name; hazardous ingredients; UN, NA, or CIS identification number; and maximum quantities stored or used on site at any time.

Hazardous Materials Management Plan (HMMP) A management and contingency plan in a form approved by the fire official that includes the following information: site plan, floor plan, material compatibility, monitoring and security methods, hazard identification, inspection procedures, employee training, and emergency equipment available.

Hazardous production material (HPM) A solid, liquid, or gas that is classified as a 3 or 4 in accordance with NFPA 704 for hazards to health, flammability, or reactivity and used in research or production where the end product is not hazardous.

Heat of combustion The amount of heat given off by a particular substance during the combustion process. A measure of fuel efficiency.

Heliport An area where helicopters take off and land from the ground or water, or from a building. Includes areas for storage, maintenance, and refueling.

Helistop Same as a heliport, except without the facilities for storage, maintenance, or refueling.

Hydrant coefficient A number describing the opening from which water flows from a fire hydrant.

Injunction A legal order issued by a court that commands a person or entity to perform a specific act or prohibits a specific action by that person or entity.

Irritant A noncorrosive chemical that has a reversible inflammatory effect on living tissue at the point of skin contact.

Label A permanent identification affixed to a product by a manufacturer indicating the function and performance characteristics of the product, name of the manufacturer, and the name of the approved testing agency that tested a representative sample of the product.

Legacy codes The codes developed by the regional organizations.

Limited area sprinkler system An automatic sprinkler system that is limited to a single fire area and consists of not more than twenty sprinklers.

Listed Equipment or materials included in a document prepared by an approved testing agency indicating that the equipment or materials were tested in accordance with an approved test protocol and found suitable for a specific use.

Magazine A structure designed and constructed for the storage of explosive materials.

Mandatory Code provisions that must be enforced and complied with.

Material Safety Data Sheet (MSDS) A document, prepared in accordance with DOL 29 CFR, that contains information regarding the physical and health hazards associated with a given product or substance and a recommended emergency action.

May Indicates a discretionary provision of the code. Enforcement is left to the judgment of the fire official.

Maximum allowable quantity Threshold quantity of a hazardous material established by the building and fire prevention codes as the maximum amount that can be stored, handled, or used within a building that is not classified as Use Group H

Model code A code developed by an organization for adoption by governments. The Uniform, Standard, and BOCA/National codes and International Codes are examples. Model codes are generally developed through the consensus process through the use of technical committees.

Mini/maxi code A code developed and adopted at the state level for either mandatory or optional enforcement by local governments, and one that cannot be amended by the local governments.

Negligence Culpable carelessness; the failure to act as a reasonable and prudent person under similar circumstances.

Nesting Method of securing compressed gas cylinders in groups, in which each cylinder has a minimum of three points of contact with other cylinders, walls, or bracing.

Net floor area Calculation used by the BNBC and SBC in determining the design occupant load of spaces with high occupancy densities, such as assembly and educational areas. Net floor area includes the area within the walls exclusive of the thickness of walls or columns or accessory spaces such as stairs, rest rooms, or mechanical rooms.

Nonsegregated storage The storage of aerosol products in general purpose warehouses within areas that are not used exclusively for the storage of aerosols.

Opening protective A rated assembly such as a door or window that provides a protected opening in a rated wall or partition. There are interior as well as exterior opening protectives.

Ordinance A law of an authorized subdivision of a state, such as a county, city, or town.

Overcrowding A condition in which the number of occupants within a building or space exceeds the approved occupant load calculated in accordance with the building code.

Performance code A code that assigns an objective to be met and establishes criteria for determining compliance. Examples are requirements for fire assemblies rated in "hours" and building materials rated as "noncombustible" when tested to specific protocols.

Permissive Code language that is discretionary and leaves application to the judgment of the enforcing authority.

Pitot pressure The pressure (in psi) shown on the bourdon gauge attached to the pitot tube inserted in a water stream flowing from a hose stream or fire hydrant.

Preaction sprinkler system An automatic sprinkler system that features dry piping, standard fusible heads, and a water supply control valve that is activated by fire detection devices. Used to reduce the possibility of water damage from accidental breakage or discharge.

Primers Explosives packages made up of an explosive charge and a detonator or detonating cord used to initiate other less sensitive explosives or blasting agents.

Protected Shielded from the effects of fire by encasement. Concrete, gypsum, and sprayed-on fire resistive coatings are all used to "protect" structural elements.

Residual pressure The pressure (in psi) measured at the test hydrant with water flowing form the flow hydrant.

Scaled distance formula A formula used to determine maximum amount of explosive material that can be detonated per delay interval of 8 milliseconds or greater, based on distance to the nearest occupied structure.

Segregated storage The storage of aerosol products within areas specifically designed for that use incorporating code-required storage arrangements and fire protection features.

Shall Indicates a positive and definitive requirement of the code that must be performed. Action is mandatory.

Signaling device Notification appliance. An alarm system component such as a bell, horn, speaker, light, or text display that provides audible, visible, or tactile output.

Smokeproof or pressurized enclosure An enclosed exit stair connected to all floors by either exterior balconies or ventilated vestibules and designed to limit the movement of smoke and fire gases into the stairwell.

Specification code A code that specifies a type of construction or materials to be used.

Spray area An area designed and constructed to be used for the application of flammable finishes complying with the requirements of the building code.

Spray booth A structure designed and constructed to be used for the application of flammable finishes, featuring power ventilation, fixed fire suppression, and separation that is installed within a building.

Spray room A room designed and constructed for the application of flammable finishes complying with the requirements of the building code.

Standard A rule for measuring or a model to be followed. How to do something, what materials to use. Also known as referenced standard.

Standard time temperature curve (STTC) Curve representing the standard reproducible test fire used since 1918 to measure the fire resistance of building materials.

Standpipe pressure-regulating device A valve permanently attached to the standpipe discharge and designed to reduce flow pressure to a predetermined level by restricting the orifice size.

Static pressure Pressure exerted within a water system at no flow.

Statute A law enacted by a state or the federal legislature.

Steiner tunnel test Test to determine the surface burning characteristics of building materials in which the flame spread of the test material is compared to asbestos cement board, rated 0, and red oak, rated 100. The higher the rating, the greater the potential hazard. In addition to flame spread, the test also measures smoke development and fuel contributed to the fire. ASTM E84, NFPA 255, UL 723,

Stemming Inert material placed in the borehole after the explosive material has been loaded to confine the effects of the explosive reaction. May also separate charges within a single borehole.

Summons A written order, issued by a judicial officer, law enforcement officer, or other authorized official, directing an alleged offender to appear in court at a specific time to answer a criminal charge. A summons issued by a law or code enforcement officer is classified as a citation.

Thermal lag The difference between the operating temperature of a fire detection device such as a sprinkler head and the actual air temperature when the device activates.

Travel distance The length of the path a building occupant must travel before reaching an exterior door or an enclosed exit stairway, exit passageway, or horizontal exit. The total length of the exit access.

Use group Building code classification system whereby buildings and structures are grouped together by use and by the characteristics of their occupants.

UST Underground storage tank for regulated liquids.

Vapor density The ratio of the weight of a given volume of a gas to that of air.

Warrant A legal writ issued by a judicial officer commanding an officer to arrest a person, seize property, or search a premises.

Water-miscible Water soluble.

Wet pipe sprinkler system An automatic sprinkler system in which the supply valves are open and the system is charged with water under supply pressure at all times.

NFPA 1031 Correlation Grid

Number	Paraphrased Job Performance Requirement*	Location
4.2.1	Prepare inspection reports, given agency policy and procedures...	2
4.2.2	Recognize the need for a permit...	2
4.2.3	Recognize the need for plan review...	2
4.2.4	Investigate common complaints...	2, 9, 10, 11
4.2.5	Identify the applicable code or standard...	1, 2
4.2.6	Participate in legal proceedings...	1, 2
4.3.1	Identify the occupancy classification of a single-use occupancy...	3
4.3.2	Compute the allowable occupant load of a single-use occupancy...	7, 11
4.3.3	Inspect means of egress elements...	7, 11
4.3.4	Verify the type of construction...	4, 5
4.3.5	Determine the operational readiness of fire suppression systems...	6, 10
4.3.6	Determine the readiness of detection and alarm systems...	6, 10
4.3.7	Determine the operational readiness of portable fire extinguishers...	6, 10
4.3.8	Recognize hazardous conditions involving equipment...	9, 13, 15, 16
4.3.9	Compare an approved plan to an existing fire protection system...	6, 10
4.3.10	Verify emergency planning and preparedness measures...	7, 11
4.3.11	Inspect emergency access	7, 11
4.3.12	Verify compliance for storage, handling, use of flammable liquids...	12, 13
4.3.13	Verify code compliance for hazardous materials...	12
4.3.14	Recognize a hazardous fire growth potential...	6, 8
4.3.15	Determine code compliance of a fire protection issue...	1, 2
4.3.16	Verify fire flows for a site...	10
5.2.1	Process a permit application...	2
5.2.2	Process a plan review application...	2
5.2.3	Investigate complex complaints...	9
5.2.4	Recommend modifications to adopted codes...	1
5.2.5	Recommend policies and procedures	1, 2, 9
5.3.1	Compute the maximum allowable occupant load...	7, 11
5.3.2	Identify the occupancy classifications of a mixed-use building...	3
5.3.3	Evaluate a building's area, height, occupancy, construction type...	4
5.3.4	Evaluate fire protection systems and equipment...	6, 10
5.3.5	Analyze the egress elements...	7, 11
5.3.6	Evaluate hazardous conditions involving equipment ...	9, 12, 13, 16
5.3.7	Evaluate emergency planning and preparedness...	7, 11
5.3.8	Verify compliance for flammable and combustible liquids and gases...	13
5.3.9	Evaluate code compliance for hazardous materials...	12
5.3.10	Determine fire growth potential in a building...	8, 9
5.3.11	Verify compliance with construction documents...	3, 4, 5, 6
5.3.12	Verify code compliance of heating, ventilation, air conditioning...	9
5.4.1	Classify the occupancy, given a set of plans...	3
5.4.2	Compute the maximum allowable occupant load...	7, 11
5.4.3	Review the proposed installation of fire protection systems...	6, 10
5.4.4	Review the installation of fire protection systems...	6, 11
5.4.5	Verify that means of egress elements are provided	7, 11
5.4.6	Verify the construction type of a building	4, 5

* Refer to NFPA 1031 for the complete Job Performance Requirements, skills and knowledge.

ABOUT THE AUTHOR

Dave Diamantes is a fire protection consultant, code trainer, curriculum developer, technical writer, and author. He retired from the Fairfax County, Virginia, Fire & Rescue Department after 25 years as Captain II, having served as a firefighter, company officer, and fire prevention officer. Since retiring from the fire department he's written extensively and worked as a consultant for state local governments. He serves on the NFPA 1031 Technical Committee on Fire Inspector and Plan Examiner Professional Qualifications. He's also the author of *Principles of Fire Prevention* (2014 Jones & Bartlett); *Report Writing for Code Inspectors* (with Dr. Jean Reynolds), and *Standpipes* (with A. Maurice Jones).

Made in the USA
Las Vegas, NV
07 July 2022